Developing innovative organizations

A roadmap to boost your innovation potential

Benoît Gailly

W0235057

First published 2011 by
PALGRAVE MACMILLAN

Palgrave Macmillan in the UK is an imprint of Macmillan Publishers Limited,
registered in England, company number 785998, of Houndmills, Basingstoke,
Hampshire RG21 6XS.

Palgrave Macmillan in the US is a division of St Martin's Press LLC,
175 Fifth Avenue, New York, NY 10010.

Palgrave Macmillan is the global academic imprint of the above companies
and has companies and representatives throughout the world.

Palgrave® and Macmillan® are registered trademarks in the United States,
the United Kingdom, Europe and other countries.

ISBN 978-1-349-33094-2   ISBN 978-0-230-29528-5 (eBook)
DOI 10.1007/978-0-230-29528-5

This book is printed on paper suitable for recycling and made from fully
managed and sustained forest sources. Logging, pulping and manufacturing
processes are expected to conform to the environmental regulations of the
country of origin.

A catalogue record for this book is available from the British Library.

A catalog record for this book is available from the Library of Congress.

Developing innovative organizations

CONTENTS

List of figures   viii

Acknowledgements   ix

About the author   x

Introduction   xi

   What the book is about   xi
   To whom the book is dedicated   xi
   What you will get out of the book   xii
   What you will not *get out of the book*   xii
   Structure   xiii

PART I: UNDERSTANDING THE INNOVATION CHALLENGE

Chapter 1   Understanding innovation   **3**

   Short case: cell phones   3
   Much more than creativity   5
   Managing change and adoption   10
   Much more than new product development   19
   How big is the change?   22
   Key points to take away from Chapter 1   23
   Further reading   24

Chapter 2   Why worry? The case for innovation **26**

   Short case: the electric car challenge   26
   Global trends   28
   Industry trends   32
   Building the right capabilities   35
   Key points to take away from Chapter 2   36
   Further reading   37

Synthesis of Part I   **39**

PART II: DEVELOPING INNOVATION CAPABILITIES

Chapter 3   Strategy: how and how much to innovate   43

Short case: strategic innovations in the
   airline industry   43
How to develop innovative corporate strategies   45
How to develop innovative business strategies   50
How innovative should your strategy be?   60
Key points to take away from Chapter 3   67
Further reading   68

Chapter 4   Nurturing entrepreneurial resources   70

Short case: boosting entrepreneurship at Plastics Ltd.   70
Innovative people: nurturing corporate entrepreneurs   72
Innovative teams: building winning packs   80
Innovative organizations: deliverables, cultures
   and structures   85
Innovative ecosystems: networks of innovation   98
Key points to take away from Chapter 4   106
Further reading   108

Chapter 5   Sourcing: finding the gold nuggets   110

Short case: identifying opportunities at Materials
   International   110
Sources of innovation   112
Harvesting corporate knowledge (internal sources)   122
Harvesting the environment (external sources)   132
Key points to take away from Chapter 5   137
Further reading   139

Chapter 6   Assessing innovations   141

Short case: raising funds for GeneticTools   141
Introduction: the art of business planning   143
Opportunity: why it is there and what it is about   147
Resources: who and how much?   158
Integrating uncertainties   167
Key points to take away from Chapter 6   174
Further reading   176

Chapter 7   Managing the innovation pipeline   **177**

   Short case: changing course   177
   Why it's not just project management   179
   Adaptative management: stages, portfolios and gates   182
   Key points to take away from Chapter 7   194
   Further reading   195

Synthesis of Part II   **196**

Conclusion   **198**

Notes and references   201
Index   206

LIST OF FIGURES

1.1	Inventiveness put to use	6
1.2	Typology of innovations	20
1.3	Incremental versus radical innovations	22
2.1	Innovation capabilities	36
3.1	Strategic value creation levers	45
3.2	Value chain	55
6.1	The business planning cycle	147
6.2	Fundamentals	161
6.3	Key valuation metrics	162

ACKNOWLEDGEMENTS

The motto of a famous scholarly search engine is "Stand on the shoulders of giants". Academic knowledge and managerial experience is indeed first built on what others have said, done or written. Similarly, this book and its content would not exist without the significant contributions of academic and executive colleagues and partners, to all of whom we are grateful.

Brilliant and experienced colleagues and peers, first at McKinsey & Company and later at the Louvain School of Management, have nourished our ideas, challenged our opinions and provided numerous illustrations and arguments.

The publications of international scholars in innovation management, quoted throughout this book, provided a second pillar upon which the book rests. They are the authors and influencers of many of the core concepts used in the following chapters, and this work would not exist without their contribution.

Third, the contribution of the numerous entrepreneurs and executives we had the chance to work with through our consulting and executive education projects cannot be stressed enough. As the frontline actors and leaders of innovation within their firms, their valuable experience, insightful opinions and challenges have greatly contributed to turning this work into a managerial handbook rather than just a theoretical dissertation.

Finally, special thanks go to Camille van Vyve, one of my former students and currently editor-in-chief of *Bizz* magazine, for her valuable feedback and multiple reviews of this text.

BENOÎT GAILLY

At present Benoît Gailly holds the post of Professor in Innovation Management and Strategy at the Centre of Research in Entrepreneurial Change and Innovative Strategies (CRECIS) of the Louvain School of Management, Université Catholique de Louvain (UCL), Belgium. He has a Master's degree in Engineering, a Ph.D. in Applied Mathematics, a Master's in Technology Studies and a Master's in Business Administration (INSEAD). His main research themes are innovation capabilities and innovation support systems, from both a policy and a business point of view.

Before working at UCL Benoit Gailly was a consultant and manager at McKinsey & Company. At UCL he held the Puilaetco Chair in Management and Financing of Technology Innovation, and was the president of the Technology Transfer Board. Benoit Gailly is also the co-director of the University Executive Education Programme in Innovation Management.

Since 1995 Benoit Gailly has been working as a board member, advisor, facilitator or expert regarding the development of innovation management capabilities for start-ups, international corporations and policy-makers. He has also been a board member of several high-technology firms, of the Belgian Federal Investment Fund, of the Brussels Regional Investment Fund and of the Technology Support Board of the Walloon Region.

His clients include international corporations in the glass, chemicals, financial, petrochemicals, ICT, pharmaceuticals, raw materials and electronics sectors and the European Commission, as well as numerous high-tech start-ups and academic spin-offs.

What the book is about

This book is about the capabilities organizations must develop if they want to grow and compete in innovation-intensive environments. Combining insights from leading academic research and experienced practitioners with concrete examples from international companies, it provides a systematic framework to understand what is innovation, why it matters, how it can be managed, and how it can help organizations to reach their objectives.

This book is not about selling one single big idea or concept as the universal solution to all innovation problems. It is about providing managers with pragmatic and rigorous structures and tools, which allow them to deal more successfully with an innovation-intensive environment.

To whom the book is dedicated

The book is dedicated to the *captains* and *coaches* of innovation. The *captains* of innovation are the men and women leading innovative projects or initiatives within their organizations, sometimes building entire new businesses out of them. They are the leaders, the entrepreneurs of innovation. They realize that the success of their project will depend upon their ability to understand the context and networks in which they operate, to mobilize and align resources, and ultimately to make change happen.

The *coaches* of innovation are the managers and consultants who aim at creating within their organization or region the conditions for innovation to strive. They are the innovation champions, facilitators and instigators. They understand that traditional "plan–do–check" approaches to management or economic development are not sufficient to compete in innovation-intensive environments, and that innovation means much more than creativity or research and development (R&D).

The book is also dedicated to management graduates and junior managers who want to enrich their management skills, experience and toolkits in order to become successful innovation captains and coaches.

Finally, the book is aimed at helping all the victims of the numerous myths that prevail regarding innovation, such as:

- innovation is *per se* a good thing, for all businesses, all the time;
- innovation means, first and foremost, being more creative and coming up with something completely new; and
- being more innovative always implies spending more on R&D and generating more new ideas.

What you will get out of the book

- A structured and systematic way to understand and tackle innovation management challenges, from a business-oriented and pragmatic point of view.
- A clear view of what innovation means from a business point of view, and why it matters.
- A roadmap to identify and develop the capabilities organizations need to manage innovation effectively.
- Key insights and tools derived from the latest academic research, consulting firms' publications and practitioners' experience.
- Concrete examples from a wide range of industries and regions.

What you will *not* get out of the book

- A survey of the latest creativity and brainstorming techniques.
- Step-by-step recipes or one-size-fits-all formulas pretending to provide universal solutions for the innovation challenges you face.

- Philosophical debates about what qualifies as an innovation and what does not.

Structure

In Part I we review what innovation means and entails from a business point of view (Chapter 1), and then discuss why and how much it should matter for firms. We conclude by outlining the capabilities firms should therefore develop (Chapter 2).

In Part II we review the core innovation capabilities firms should develop and maintain in order to manage innovation effectively and turn it into a source of competitive advantage. We first discuss the strategic role of innovation (Chapter 3) and the importance for the organization to build entrepreneurial resources (Chapter 4). We then detail how firms can develop and improve the way they proactively identify a continuous flow of opportunities (Chapter 5), the way they systematically assess which opportunities would create a balanced portfolio (Chapter 6), and how they can capture those opportunities in a dynamic way (Chapter 7).

The table presented below outlines the key issues managers dealing with innovation must manage (first column). We map these with the key concepts (second column) and selected innovation tools (third column) presented in the book.

What innovation is (Chapter 1)

Making sense of it	More than creativity More than inventions	*Definition of innovation* *Newness is relative*
Managing change	Managing diffusion Managing disruption	*Factors of adoption* *Disruptive innovation*
Defining the scope	Delivering new offers and in new ways Managing small and big changes	*Product and process* *innovations* *Incremental and radical* *innovations*

Why it matters (Chapter 2)

Understanding why it matters	Analyzing megatrends	*The new economy*
	Analyzing industry trends	*New challenges*
Analyzing what it means for organizations	Fostering the entrepreneurial process	*Scan, assess and implement*
	Steering the entrepreneurial process	*Strategic vision and entrepreneurial resources*

Innovation strategy (Chapter 3)

Designing a corporate innovation strategy	Defining the corporate scope	*Exploration versus exploitation*
	Defining how much to innovate	*Drivers of innovation strategy*
Designing a business innovation strategy	Defining innovative value propositions	*First mover (dis)advantages*
	Defining innovative value chains	*Strategic innovation*
	Aligning strategy and innovation	*Innovation postures*

Entrepreneurial resources (Chapter 4)

Nurturing entrepreneurs	Identifying entrepreneurial traits	*Role models*
	Analyzing entrepreneurial behavior	*Entrepreneurial activities*
	Fostering entrepreneurship	*Drivers of intention*
Managing innovative teams	Setting up innovative teams	*Handling diversity*
	Developing innovative managers	*Stretch versus stress*
Building innovative organizations	Understanding innovation metrics	*New mindsets and new businesses*
	Fostering an innovation culture	*Entrepreneurial orientation*
	Dealing with size	*Ambidextrous organizations*
	Designing innovative structures	*Corporate venturing*
Managing innovative ecosystems	Understanding the innovation systems	*Innovation support mechanisms*
	Leveraging innovation networks	*Open innovation*

Proactive deal-flow (Chapter 5)

Exploiting the sources of innovations	Understanding that R&D is not enough	*Sources of innovation*
	Combining technologies and needs	*Push and pull*
Harvesting corporate knowledge	Fostering organizational learning	*Gatekeepers*
	Handling knowledge management	*Absorptive capacity*
	Managing intellectual property rights	*Pros and cons of patents*
Harvesting external opportunities	Harvesting the market	*Value gap analysis*
	Harvesting entrepreneurs	*Corporate venture capital*

Balanced portfolio (Chapter 6)

Assessing innovation	Business planning	*Business planning cycle*
	Assessing why (strategy)	*Sources of differentiation*
	Assessing what (business model)	*Offer and value chain design*
	Assessing who (human resources)	*Governance structure*
	Assessing how much (finance)	*Key value drivers*
	Valuing an opportunity	*Net present value*
Integrating uncertainties	Managing known unknowns	*Sources of uncertainties*
	Running sensitivity analysis	*Monte Carlo simulation*
	Running scenario analysis	*What you need to believe*

Nimble execution (Chapter 7)

Managing innovation projects	Building an ambidextrous organization	*Effectuation*
	Managing the portfolio	*Balanced portfolio*
	Handling adaptative management	*Stage-gate process*
Implementing innovation process	Managing stages	*Stage examples*
	Managing gates	*Gate examples*
	Dealing with decision biases	*Why we get things wrong*

UNDERSTANDING THE INNOVATION CHALLENGE

Innovation means many things to many people. Moreover, the fact that it is important is nowadays often taken for granted. But innovations vary in terms of scope, pace, complexity or risk. What is seen as business as usual in one organization may be perceived as being very innovative in another. Similarly, while we shall argue throughout this book that innovation matters for all organizations, it is important to clarify how much it matters for a given firm, and why.

As a consequence, the aim of this first part is to set the scene by addressing two issues: what is innovation from a business point of view; and why it matters for organizations. In Chapter 1 we shall first discuss why innovation is much more than just creativity and/or inventiveness; and then, in Chapter 2, highlight why innovation should be on the agenda of all organizations, whether low- or hi-tech, small or large.

CHAPTER 1

Understanding innovation

Short case: cell phones

Mobile telephony is actually quite an old idea. The concept of being able to communicate without wires goes back to the first radio phones developed before the Second World War. The first automated mobile phone service was developed by Ericsson in the 1950s, and the first modern cell phone was sold by Motorola in the 1970s. Both firms then developed many new products, in particular sophisticated handsets, based on their technology and engineering skills.

In parallel, progresses in electronics and computing led over time to the emergence of more sophisticated handsets and more powerful communication standards (Bluetooth, Wi-Fi, 2G, 3G and now 4G). This has allowed the delivery of new types of content beyond voice calls (such as text, images, data, software or videos) and the addition of new features on mobile handsets (electronic agenda, camera, music player, GPS, television and so on).

But this linear invention-based and technology-driven summary of the history of cell phones hides a much more complex and chaotic process.

First, the development of cell phone services had to overcome two large hurdles: no market and no compatible infrastructure. Indeed, the market studies completed in the 1980s consistently predicted that the global market for cell phones was marginal (a few million customers worldwide at most). In other words, there was a priori no visible demand.

→

Furthermore, the small potential market identified added to the risk of cannibalization of existing services could not justify the extensive infrastructure investments required.

It was only when several European countries agreed in the 1990s on a common standard (the Global System for Mobile Communications, or GSM) and when people began to observe and experiment with the potential use of cell phones that the market really took off.

Two other non-technological innovations played an important part:

(i) The development of text messaging services, whose initial success was a complete surprise to most operators, and which now represent a significant share of revenue for many of them.

(ii) The development of new business models based on pre-paid fees rather than the standard subscription-based model used for traditional, fixed-line services. These new approaches allowed cell phones to be sold and bought in a box. They also allowed users (and parents) to control monthly bills. In some markets, more than 50 percent of customers use those "pre-paid" offers.

Second, the adoption of cell phones has been hindered throughout its development by the perceived complexity of the handsets (multiple menus and features, complex invoices, or fears about damage to the user caused by electromagnetic radiation). In particular, early personal digital assistants and smart phones developed by Apple (the Newton) and Phillips (the Synergy) were failures. Conversely, Nokia based its initial success largely on the simplicity, ease of use and convenience of its handsets. In particular, users were able to switch from one Nokia model to another without reading complex user manuals.

Third, while the industry has since 2000 invested billions in the development of high-technology standards, it is once

$\rightarrow$

again disruptive approaches that have shaken the industry. These include new value propositions based, among others, on simplicity and ease of use (such as the touch screen, new operating systems or longer battery life) and on new business models (such as the application platforms or pricing based on flat rates).

In summary, while the cell phone industry is highly technology-intensive and has seen many new concepts emerge:

- major innovations were not R&D driven;
- the adoption of innovation was driven more by norms and standards, ease of use and convenience than by technological sophistication; and
- the early inventors were not necessarily the successful innovators.

In other words, producing creative ideas and investing heavily in R&D in order to be the first to develop new sophisticated products clearly does not guarantee success. Indeed, managing innovation means much more than that.

Much more than creativity

In the mythology of innovation, a lonely hero (named Newton, Einstein or Darwin) comes up suddenly with a great idea ("the eureka moment"). That great idea is then immediately recognized by everybody around him (it is generally a "him" in the myths) and becomes an instant success. The hero becomes rich and famous, and is remembered as a genius. In such myths, innovation is mainly a question of creativity and of coming up with great ideas. But as all the firms who tried to become innovative by organizing brainstorming sessions or setting up idea portals have experienced, this is indeed a myth.

History shows us that the same concept can be a complete failure in one market and a success in another, apparently similar, one. And a complete failure can become a success just a few years later. The same idea can be launched successively and abandoned several times in the same industry.

Solving those puzzles requires the myths to be dispelled and to understand what innovation is about. In this chapter, we shall review how innovation can be defined from a business point of view; how it is linked strongly to change; how it means, therefore, much more than creativity; and why whether an idea is new is a relative concept.

Definition and implications

One can find many different definitions of innovation, from the simple "inventiveness put to use"[1] to the complex "the development and implementation of new ideas by people who over time engage in transactions with others within an institutional order".[2] Whether you are a sociologist, an economist, a journalist or an historian will change how you define innovation.

From a managerial point of view, we can define innovation as the combination of two key features: newness (invention or discovery) and change (see Figure 1.1).

An innovation is obviously something that is new for an organization. It can result from an invention (something new was made) or a discovery (something new was found). It is driven by the technological capabilities of the organization, by the ideas it can handle and by its knowledge base.

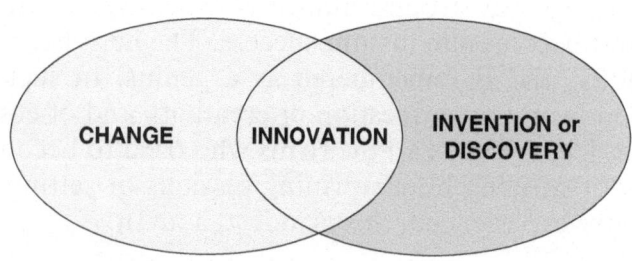

Figure 1.1   Inventiveness put to use

But an innovation is more than an invention or a discovery. It is something that generates a change and that has an impact on the market structure and on the social context of an organization. It is therefore also driven by the know-how of the organization and its ability to mobilize resources and modify its structure, style, systems, procedures or values. It leads the organization to reach a new state.

As a consequence, the ability of an organization to manage innovation successfully will be as much driven by its ability to handle newness than by its ability to make change happen. While the former is already difficult (decades and billions have been spent on some inventions such as nuclear fusion or treatments for HIV), it is the latter with which most firms struggle.

Hence invention is about seeing new things or seeing the world differently; about being convinced by something new. But innovation is about *doing* new things or doing things differently; about being convincing about something new.

This has three important implications that will be discussed below:

- whether something is new is relative;
- managing innovation is not managing inventions; and
- for most organizations, innovation is not a problem of (lack of) creativity.

What is new?

A change will have to be managed as an innovation by an organization if it leads to a state that is perceived as being new *within that organization*, in terms of market or technical capabilities, or in terms of component or architectural knowledge. This is, for example, the case when a firm develops new solutions for existing problems, such as new ways to generate electricity or heat using renewable energy sources.

But a change will also have to be managed as an innovation if it leads to a state that is perceived as being new *within the environment* (market or industry) where the firm operates. This is, for example, the case when a firm implements novel combinations of existing technologies, such as financial institutions developing online services.

Finally, the change can be perceived as new *both within the firm and its environment*. This is the case when technology and markets co-evolve, such as with the development of space tourism, which is both a new market and a technology challenge.

Let us stress that this means that whether an idea is in fact really new regarding the state of the art (nobody thought of it before) does not really matter from an innovation management point of view. What matters is that it is perceived as new by the organization, by its environment, or both, and how much "liability"[3] this newness generates. Hence what is new is relative.

More than inventions

An invention or a discovery can be defined as the finding of a concept, idea, method, process, technique or prototype, or a combination of these, that was not known before. It is about creative thinking, coming up with something new. Moreover, "as long as they are not carried into practice, inventions are economically irrelevant. And to carry any improvement into effect is a task entirely different from the inventing of it, and a task, moreover, requiring entirely different kinds of aptitudes" (Joseph Schumpeter[4]).

Invention means, therefore, imagining new things, defined as novelties *vis-à-vis* the current state of the art (see above) and can usually be associated with the work of individual geniuses (Einstein, Pasteur, Darwin, for example). An invention can also be seen like a final destination resulting from a linear search process ("Eureka, I have finally found it!"). You can date an invention (for example, the electron was discovered in 1900) and the discontinuity in knowledge it created.

In contrast, an innovation refers to the actual implementation of a value proposition having no close substitute in a given context. It is about doing new things intentionally; getting new things done.

Because it involves change and implementation, an innovation is about realizing new combinations rather than finding new individual things. While the inventors discover missing pieces, the innovators are the ones who construct new "puzzles"[5] out of them.

Furthermore, innovations are realized by entrepreneurs and the organizations they build, not by isolated individuals. And finally, the development of an innovation is a continuous journey, with a strong path dependency, rather than a point in time, a destination.

The rise of global telecommunications, the emergence of the personal computer and of e-commerce, or even the development of the theory of evolution, provide examples of innovations that meant much more than mere inventions. They emerged from the combination of many new and old ideas, involved substantial social, technical and economic changes, and were driven by multiple organizations and entrepreneurs.

Hence inventions and innovations refer to very different concepts, which need to be managed in different ways.

From an innovation management point of view, this means that the challenge for an organization is not to be inventive – in other words, to lead individuals to reach destinations that push the limits of the state of the art. The challenge is rather to be innovative – in other words, to manage a continuous process where people implement combinations that are new in their context.

As an illustration of this difference, let us remember that Kodak invented the technology of digital photography (in 1975!) but it was Sony and others who were successful at managing the resulting innovations. History is indeed replete with "cases in which the inventor of major technological advances fails to reap the profits from his breakthroughs".[6]

More than creativity

One important consequence of the differences between inventions and innovations, and of the relativity of newness, is that, from an innovation management perspective, creativity is in most cases not the main issue. While there is evidence that creativity techniques can be very powerful when attempting to find alternative solutions to known problems, there is also plenty of evidence that more creativity does not necessary lead to better innovation.

In other words, managing an innovation effectively might involve a dose of creativity, but it means much more than that. It is about making change happen. When senior executives are asked "Which are the main obstacles to innovation in your firm?"[7], "Not enough great ideas" appears far behind "Risk-averse culture", "Lengthy development process", "Lack of coordination" or "Difficulties in selecting the right ideas".

To paraphrase Thomas Edison, probably one of the greatest innovators, we can say that innovation is "1 per cent inspiration and 99 per cent perspiration". More recently, what firms such as General Electric (GE) or even Apple see as the main success factor of their management of innovation is not creativity but operational excellence, expertise and professionalism.

Managing the change involved in an innovation means understanding why some users are resistant to that change while others will quickly switch. It is about understanding that the less you disrupt people, their systems, their routines and their values, then the faster and more easily will they accept something new to them.

Managing change and adoption

"There is nothing more difficult to take in hand, more perilous to conduct, or more uncertain in its success, than to take the lead in the introduction of a new order of things" (Niccolò Machiavelli, *The Prince*, 1537).

Most people show a certain aversion toward change when that change involves something new to them. Their instinct is to resist abandoning the way they have always done things. History is full of examples of new technologies that raised fears (telegraph wires that could affect the weather, or trains that would generate nervous disorders) that affected their diffusion. Even Socrates reviled the idea of writing texts (rather than memorizing ideas), and media innovations such as novels, cinema or comics books were all once considered dangerous. More recently, innovations such as nuclear energy, wireless telecommunications (and the electromagnetic fields they generate) or genetically modified organisms see their development influenced by people's perceptions, beliefs and fears.

While such fears or resistance might or might not be justified, the way they are dealt with will have a critical effect on the success of an innovation. This involves dealing with fears that are often impossible to disprove (such as the controversy regarding the side-effects of child vaccines). It also means balancing the risks of an innovation with the benefits of legitimate uses (for example, in the case of online privacy issues). Finally, it means finding and promoting pragmatic trade-offs between precautionary principles and technocratic approaches.

In particular, it also involves identifying and dealing with those who will lose out from the adoption of an innovation. To paraphrase Gary Hamel: "Management innovation often redistributes power so [one should not] expect everyone to be enthusiastic."[8]

The diffusion challenge: disrupt competitors, not users

Managing the change aspect of an innovation successfully requires an understanding of the drivers of diffusion – the factors that influence why some innovations quickly reach a high adoption rate while others, apparently of similar value, do not. As an example, cell phones were adopted all over the planet within about a decade, while barcodes, invented in the 1970s, are still not used in many sectors or regions.

The key issue is whether, for a given innovation, the perceived added value brought by newness is more important than the perceived disruption generated by the change, and how, as a consequence, a firm can manage both. While maximizing the perceived added value is probably already on the marketing agenda of most firms, how to minimize disruption is often neglected.

The main drivers of diffusion of an innovation have been identified by Rogers.[9] While his work focused initially on consumer goods, it provides a good general framework for various types of innovation.

The first factor identified by Rogers relates to the benefits of *newness*: the greater the perceived advantage an innovation brings relative to what existed before, the faster the diffusion. Hence the fast diffusion of innovations that bring direct tangible benefits such as comfort or speed, and the importance of new features and "coolness" factors.

Innovations that struggled with this factor include new digital media technologies (DVD first, then Blu-ray), where the perceived benefits *vis-à-vis*, respectively, videotapes and video-on-demand or streaming were initially not clear for many potential customers. Another example is high-performance double-glazing, where the insulation benefits provided are not directly visible to the user, though the price to be paid is.

Rogers also identified four other factors, all related to the *change* aspect of an innovation; in other words, to the level of perceived disruption for users. What matters is to convince potential users and partners that the innovation is not threatening and can easily and advantageously be integrated into routine and daily practice. Those four factors are detailed below.[10]

First, the perceived disruption will be lower (and therefore the diffusion easier) if the innovation is highly compatible with existing uses, values and norms. The higher the compatibility, the lower the switching costs, be it in terms of tangible costs (physical investments) or more tacit barriers (change of habits, training required, cultural distance, or traditions).

The electronic purse is an example of an innovation whose diffusion is being slowed by this aspect. Indeed, it requires the use of new equipment (such as card readers at points of sales) and it challenges the traditional notions of what is a wallet (a pocket with coins and notes) and what is a bank (where people's money sits). It also challenges the idea that cash remains anonymous.

Second, the perceived disruption will be higher if the innovation is more complex and if it involves dealing with a large volume of information or with perceived uncertainties (will it work?). The issue here is whether the added value is directly transparent to the user and whether the decision-making process involved is simple (such as the one-click purchase process on Amazon).

Examples of innovations strongly affected by this second factor include those in consumer electronics, such as digital music players or cell phones, which remain complex to use for many people, or software and its often cryptic jargon.

Third, users will be more likely to take the risk of disruption if their action is reversible – in other words, if they can try before they buy. If you provide users with the opportunity to experiment with an innovation, you can expect that many will try it and more will adopt it. Retailers with their money-back guarantees as well as all the firms investing in showrooms have for a long time understood this aspect.

The last factor is related to social pressures and conformism: a user will be more likely to adopt an innovation and will perceive the disruption to be less risky and more attractive if he or she sees others around him or her doing the same. But this is possible only if the innovation is observable; in other words, if the adoption by others is visible.

Celebrity endorsement and recognizable trademarks (logos) on cars such as "turbo" or "hybrid" increase observability. In contrast, many innovations inside a car (such as new materials used, or new electronic systems installed) remain invisible. Similarly, a new generation of televisions (which often sit in the living room) will diffuse faster than a new generation of washing machines (which people do not see when they make a social call).

This means that the diffusion of an innovation is driven by the perceived disruption as much by (and often more than) the perceived added value. Moreover, the key factors identified above should be analyzed a priori (to assess the likelihood of rapid diffusion) and dealt with proactively (they are not frozen) when managing an innovation.

Managing diffusion: fertile ground and cultural ambience

The diffusion of an innovation will be affected strongly by the characteristics of the innovation itself (see above). But the behavior of the stakeholders involved, in particular the provider and the potential users of an innovation, also have a strong influence. In particular:[11]

- whether the provision and adoption of the innovation involves individual, collective or centralized decision processes (for example, the purchase of a new toy for a child for Christmas or the investment in a new information technology system by a firm);
- what are the communication channels involved, both in terms of mass media and interpersonal relations (for example, the adoption of social networking tools);
- what are the norms and degree of interconnection of the group of users (for example, the adoption of new energy drinks by young people);
- what are the promotion efforts launched by the change agents (advertisers, development agencies) and the competitive intensity (such as in the case of new "bio" products); and
- what are the network effects involved, if any.

Indeed, "major innovation decisions are largely a political process, often involving professional groups advocating self-interested outcomes under conditions of uncertainty (ignorance), rather than balanced and careful estimates of costs, benefits and measurable risk".[12]

Finally, the "cultural ambience"[13] or the wider social context in which the innovation emerges also plays a significant role. In particular:

- The manufacturing organizations and network of firms and authorities devoted (or enlisted) to the continuous maintenance of the technology and its sub-elements (intermediaries, complements). For example, a whole ecosystem of firms and market mechanisms had to be created to support the development of 3D television.

- Whether the innovation can acquire meaning (social constructs) and connotation (be integrated in discourse) to reach public acceptance. For example, advertising a car as "green" would have not meant anything in 1990. Similarly, being "chemical" in the 1950s or "electronic" in the 1980s was seen as a quality.

- Whether the innovation can find timely political and legal acceptance, in particular regarding product regulation – for example, in the telecommunication sector (where the use of wireless spectrum is regulated) and in the media sector (where copyright and censorship issues are common). Similarly, renewable energy sources have much higher political and legal acceptance in 2010 than they had in 2000.

The diffusion process

We discussed above the factors that influenced the diffusion and adoption of an innovation, related to the innovation itself, the stakeholders involved and the wider social context. Another important aspect of the diffusion process is that different sets of users and different versions of the same innovation will be involved over time.

As an example, not all firms are expected quickly to adopt the new RFID (radio frequency identification) technology, intended to replace the barcodes used in many industries. From large retailers forcing the adoption of this technology throughout their supply chain to niche small and medium-sized industries (SMEs) with limited exposure to computing, various segments are likely to emerge that will or will not adopt this technology over time.

Moreover, faced with this innovation, not all users will react in the same way and at the same time. It is, for example, common to group users into five different categories, corresponding to five different stages of evolution of the adoption of an innovation.[14]

First, the innovators, the adventurous or "geek" users, who sometimes like new things just because they are new and who do not perceive the change involved in very negative terms. They correspond to an *introduction stage.*

Second, the early adopters, the visionary trendsetters, who find and define where the market or the industry is going. They correspond to *the growth stage of adoption.*

Together, these two segments represent only a minor share of the potential market and tend to correspond to more interactive and open-minded users. They are often better-educated but not necessarily younger than the average user.

A common mistake at this stage is for firms to consider that, because this first part of the market was converted relatively rapidly, the rest will follow easily. As a consequence, the firms decrease their focus on the diffusion of their innovation –for example, by switching too quickly to newer projects. They forget, at their peril, the "chasm"[15] that often exists between the innovators and early adopters and the rest of the market.

Next come both the *early majority* and *late majority* segments, of more thoughtful, pragmatic or even skeptical customers who represent the bulk of the market. They correspond respectively to a turbulence and maturity phase of the adoption, where the adoption rate first marks an inflexion and then slowly decreases. Finally come *the laggards,* the more conservative or traditional customers who are converted often only when the market has already reached a stage of *decline.*

While these five segments and stages are probably too deterministic and have not always been observed empirically, they highlight the importance of identifying the diversity of customers' attitudes toward an innovation. This implies a need to understand who can be converted when, and to adjust the firm's strategy accordingly.

In the same way that the target users evolve over time, the innovation itself will be subject to a succession of adjustments, modifications and bifurcations as firms compete, markets evolve and technology matures. Only after some time (sometimes decades) will a "dominant design"[16] emerge, defining how the product is supposed to look and operate in the mind of users, and the implicit requirements that will form the basis for copy-cat models. At that stage a *de facto* standard appears, corresponding to a tipping point, a consolidation where firms switch their focus from *what* to offer to *how* to offer it (see next section), or exit the market entirely.

The emergence of such dominant design will in particular be influenced by the adoption of formal regulations and standards, by the existence of co-specialized assets and switching costs, and by the strategic maneuvering of the firms involved.

For example after the Second World War there were still firms developing cars with six or eight wheels rather than the now standard four. Similarly, it took several years for modern wind turbines to converge to the three arms facing the wind design now commonly in use.

Disruptive innovation

"The perfect is the enemy of the good" (Voltaire).

One way to exploit the importance of the change aspect of innovation discussed above is to turn the concept on its head. Rather than to try to come up with something very new, very advanced, very sophisticated and then try to convince users to change, some firms focus on finding what will make it easier for users to change in the first place. Rather than focusing on trend-setting early adopters, they target the change-averse late majority.

Such "disruptive innovations"[17] correspond to offers that are not as sophisticated as those already used in established markets (they often offer less functionality), but that have other attributes – most often simplicity, convenience and low cost – that appeal to a new, small and initially unattractive (to established firms) set of customers, who use them in new or low-end applications.

Disruptive innovations tend to enable less-skilled or less wealthy customers to do for themselves things that only the wealthy or skilled intermediaries could do previously. They compete against existing offers, not because they perform much better but because they are more convenient to use.

For example, Siemens wants its business to develop offers that are "SMART" (Simple, Maintenance friendly, Affordable, Reliable and robust, and Timely). Another example of disruptive innovation is the digital camera, which was initially less sophisticated than the best existing film-based cameras. It was therefore initially ignored by professional photographers but provided simplicity and ease of use for basic users. As a consequence, incumbent firms such as Kodak or Agfa somewhat failed to see the threat coming and probably reacted much too late to this innovation.

Opportunities for disruptive innovations emerge in particular when industries "overshoot".[18] In such situations, competing incumbents add features at a pace faster than the speed at which most customers can utilize or absorb them. They focus on high-end customers and improve product performance and functionality beyond what most customers want, often at the expense of price.

In these situations, incumbents are vulnerable to a new entrant targeting the mass market with a simpler offer. As an example, most financial institutions were slow to react to the emergence of online stock trading services, perceived as being too basic compared to the supposed sophistication and added value of their traditional offers.

Disruptive innovations are particularly tricky to handle because they counter the traditional strategic focus to offer the best products to the firm's best customers. Indeed, incumbents will often focus on their market share and product margin in the most attractive market segments, where performance is well understood and can be monitored. In contrast, disruptive innovators will develop cheaper and more convenient offers and will be ready to focus on a priori less profitable and/or emerging markets in order to fill the vacuum left by the incumbents' focus on performance.

Let us stress that it remains a risky strategy, as the multiple failed attempts to introduce and sell a simple computer (given that most people do not need the level of performance and functionalities offered by incumbents) have shown.

Much more than new product development

Just as innovation cannot be restricted to new ideas or inventions, neither can it be restricted only to new products. From new ways to communicate to new forms of organization, from adjustments to existing offers to completely new businesses, there is a wide variety of innovations a firm must be able to handle.

Below we shall develop a pragmatic but systematic typology of innovations highlighting this diversity.

A definition-based typology of innovations

As we highlighted in the previous section, an innovation has two important facets: newness and change. Innovations will therefore differ in terms of (i) what is new (the object); and (ii) how big is the change (the intensity).

The first way we can categorize innovation is therefore by differentiating what is new: the offer or the way it is offered. The second way to differentiate innovations is by assessing the importance of the change involved, from small (incremental) to large (revolutionary). Here we shall describe the four categories emerging from these two dimensions (see Figure 1.2).

Delivering new offers

New offers can relate to new products or new services available on the market, or new ways to offer those products and services (new business models). The OECD indeed defines product innovation as "the introduction of a good or service that is new or significantly improved with respect to its characteristics or intended uses".[19] It is therefore about what the firm sells, about its outputs. This includes:

- designing new or improved products and services, or offering new interfaces (for example, with new cell phones);

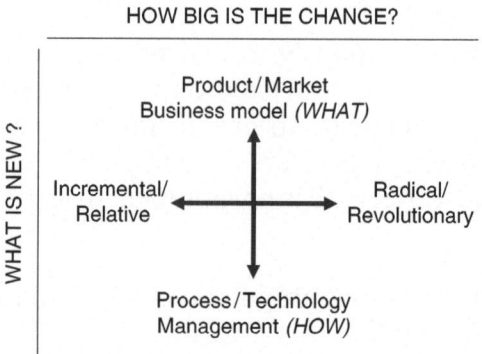

Figure 1.2   Typology of innovations

- developing marketing innovations, related to pricing, packaging, delivery, promotion or placement (for example, new ways to sell home-made coffee or beer);
- targeting new market segments (for example, providing financial services to niche markets);
- developing whole new business models (for example, virtual worlds such as Second Life or social networks such as Facebook); and
- any combinations of the above.

Let us stress here the importance of service innovations, which have historically often been neglected. Service innovations might include innovations involving basic technologies and/or low technical skills (such as in cleaning or hairdressing), or areas characterized by high professional qualifications and advanced use of computing (such as in financial services or consulting).

Moreover, as services are by nature intangible and involve interactions, service innovations are often difficult to protect (for example, through patents) and industrialize. Finally, the increasing role of information in many service industries has paved the way for significant information-technology-based innovation – for example, in retailing and logistics.

Delivering or producing offers in new ways

New ways to deliver an offer relate to all the processes in the value chain of an industry that can be improved or changed. From definitions focusing initially on manufacturing innovations, the OECD now defines process innovations as "new elements introduced into an organization's production or service operations: input materials, task specifications, work and information flow mechanisms and equipment used to produce a product or render a service".[20] It is therefore about how the firm operates and with whom; about its effectiveness. This includes:

- new delivery methods, from new R&D processes to new manufacturing techniques and new ways to distribute a product (for example, in the biotechnology industry);
- organizational innovations: new business practices and enabling technologies, new workplace organization and human resource management, new types of external relations and partnerships (for example, in the auto industry);
- new sources of competitive advantage (such as low-cost or disintermediation strategies); and
- any combinations of the above.

Let us stress that the two categories highlighted above (what/ product and how/process) are strongly interdependent. On the one hand, most product innovations involve some level of process innovation. Reciprocally, most process innovation will affect the end-product to some degree, though many process innovations are not visible to the end customers.

On the other hand, product innovations tend to prevail in the early stages of the development of an innovation (the "fluid/ ferment stage"[21]), when firms experiment with alternative designs, while process innovations will be more important at later stages, once a dominant design has emerged (see above) and firms focus on cost reduction and/or customization.

Finally, at the level of the economy, the product of one firm (or industry) is often used to produce goods and services in another. As a consequence, a product innovation by the former will be a process innovation for the latter (for example, in the steel and the construction industries).

How big is the change?

The second important way to differentiate innovations is related to the importance of the changes involved. The issue is whether the innovation is about improving an existing product or service to increase competitiveness (incremental) or, at the other extreme, is about launching major or structural changes that will make existing products or services obsolescent (radical innovation) (see Figure 1.3).

Incremental innovations will not affect the nature of the business, where the firm operates or how it is organized. It relates to exploitation and continuous reinvention of a business (almost) as usual and through marginal improvements. It is about enhancing existing or familiar technology capabilities as well as leveraging market knowledge to capture new opportunities that fit with the current strategy and/or business model. As an example, fast-moving consumer goods (FMCG) companies continuously tinker with their products and value chain without fundamentally changing the nature of their activities.

On the other hand, radical innovations are about exploring new businesses, new forms of organization and/or new markets. They

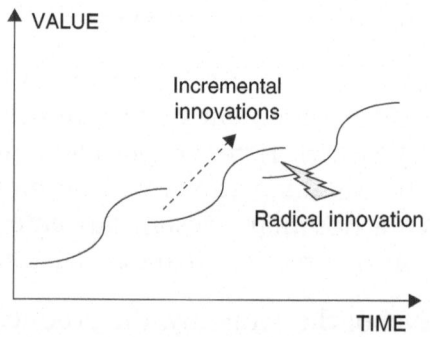

Figure 1.3   Incremental versus radical innovations

are about developing new knowledge bases and new technologies in order to reconfigure the market and build new strategic options. As an example, most media and music firms are exploring new ways to make money in a virtual world.

Obviously this is not a black/white categorization. One can find in each industry innovations that span all the way from minimal adjustments to revolutions. Yet two key issues are worth highlighting here.

First, it is the degree and difficulty of the change involved that matters from an innovation management point of view, not the sophistication of the underlying technology.

As an example, switching from cathode to plasma and LCD televisions (and then LED and perhaps OLED screens) has involved a huge technological challenge for the firms involved. But from the point of view of the user, the successive television sets they buy are often seen mainly just better looking and flatter. The business model of broadcasting and (unfortunately?) most of the content delivered is still pretty much the same as it was in the 1960s.

In contrast, the aircrafts that low-cost operators use are pretty similar to those used by traditional airlines (involving the same underlying technology), but these companies revolutionized air travel (they generated radical innovations).

Second, to decide how long to continue to improve (or exploit) an existing business model, and when and how much to switch to (or explore) a new generation is probably one of the toughest strategic decisions for a firm to take, where there is in most cases no (without insights) easy answer. In other words, there is often a very thin line between whether a manager is a visionary genius or a stubborn conservative. We shall return to this issue in Chapter 2.

Key points to take away from Chapter 1

Managing innovation means much more than boosting creativity or inventiveness; the lack of good ideas is in most cases not an issue. Managing innovation is about managing change and adoption. In particular, the drivers of adoption can and should therefore be analyzed and managed at product, firm and social context levels.

Technological sophistication does not mean added value. In particular, disruptive innovations can threaten incumbents, through change-friendly features such as ease of use and convenience.

Innovation is not only about new products. It is both about what you do (product, service, business model) and how you do it (value chain, organization, strategy).

Finally innovation is about how much change you (can) generate; about combining both incremental/exploitation and radical/exploration opportunities.

Further reading

Afuah, A. (2003), "Static Models of Innovation" in *Innovation Management: Strategies, Implementation, and Profits*, 2nd edn, Oxford University Press, pp. 14–32.

Amit, R. and Schoemaker, J. H. (1993), "Strategic Assets and Organizational Rent", *Strategic Management Journal*, Vol. 14, pp. 33–46.

Anderson, P. and Tushman, M. (1990), "Technological Discontinuities and Dominant Design", *Administrative Science Quarterly*, Vol. 35, pp. 604–634.

BCG (Boston Consulting Group) (2009), *Innovation 2009: Making Hard Decisions in the Downturn* (Boston, MA, BCG).

Berkun, S. (2007), *The Myths of Innovation*, Sebastopol, CA, O'Reilly Media.

Christensen, C. (1997), *The Innovator's Dilemma: When New Technologies Cause Great Firms to Fail* (Boston, MA, Harvard Business School Press).

Christensen, C. M. and Raynor, M. E. (2003), *The Innovator's Solution* (Boston, MA, Harvard Business School Press).

Davila, T., Epstein, M. J. and Shelton, R. (2006), *Making Innovation Work* (Philadelphia, PA, Wharton School Publishing), pp. 29–58.

Dess, G. and Lumpkin, G. T. (2003), *Strategic Management* (Boston, MA, McGraw-Hill/Irwin), Ch. 12.

Evans, H. (2004), *They Made America. From the Steam Engine to the Search Engine. Two Centuries of Innovators*, New York, Back Bay Books.

Fagerberg, J. (2005), *The Oxford Handbook of Innovation* (Oxford University Press), pp. 433–481.

Gollier, C. (2001), "Should We Beware of the Precautionary Principle?", *Economic Policy*, Vol. 33, pp. 303–327.

Henderson, R. M. and Clark, K. B. (1990), "Architectural Innovation: The Reconfiguration of Existing Product Technologies and the Failures of Established Firms", *Administrative Sciences Quarterly,* Vol. 35, pp. 9–30.

Ireland, R. D., Covin, J. G. and Kuratko, D. F. (2009), "Conceptualizing Corporate Entrepreneurship Strategy", *Entrepreneurship Theory & Practice*, Vol. 33, pp. 19–46.

Moore, G. A. (2002), *Crossing the Chasm: Marketing and Selling High-Tech Products to Mainstream Customers*, New York, Harper Paperbacks.

Nelson, R. R. and Winter, S. G. (1982), *An Evolutionary Theory of Economic Change*, Cambridge, MA, Harvard University Press.

OECD/Eurostat (2005), *Guidelines for Collecting and Interpreting Innovation Data – The Oslo Manual*, 3rd edn, Paris, OECD.

Rogers, E. M. (1995), *Diffusion of Innovations*, New York, Free Press.

Schilling, M. A. (2006), *Strategic Management of Technological Innovation*, 2nd edn, Boston, MA, McGraw-Hill, pp. 63–71.

Schumpeter, J. A. (1939), *Business Cycles: A Theoretical, Historical and Statistical Analysis*, Boston, MA, McGraw-Hill.

Smith, D. (2006), *Exploring Innovation*, Boston, MA, McGraw-Hill, pp. 22–42.

Tidd, J., Bessant, D. and Pavitt, K. (2001), *Managing Innovation,* 2nd edn, Hoboken, NJ, John Wiley, pp. 6–13.

Tushman, M. L. and Anderson, P. (1988), "Technological Discontinuities and Organizational Environments", *Administrative Sciences Quarterly,* Vol. 31, pp. 439–465.

Tushman, M. L. and O'Reilly, C. A. (2002), *Winning through Innovation*, Boston, MA, Harvard Business School Press, Ch. 7.

Utterback, J. M. (2005), *Mastering the Dynamics of Innovation*, Boston, MA, Harvard Business School Press.

Van de Ven, A. H. (1986), "Central Problems in the Management of Innovation", *Management Science*, Vol. 32, pp. 590–607.

Van de Ven, A. H., Polley, D. E., Garud, R. and Venkataraman, S. (2008), *The Innovation Journey* (Oxford University Press).

CHAPTER 2

Why worry? The case for innovation

Short case: the electric car challenge

Developing electric cars is actually an old idea. At the end of the nineteenth century most speed records for cars were held by electric ones, and electric car sales peaked just before the First World War. Infrastructure and public perception problems, combined with new combustion-engines technologies (less noisy, easier to start and more powerful) and increased availability of cheap oil then led the combustion engine to become the focus of car companies' strategy.

More recently, following threats from new rules regarding zero-emission vehicles, in 1996 General Motors (GM) developed a fully functioning electric car (the EV1) which was sold in California. While there were apparently customers queuing to buy the car, the project was scrapped after a few years, among other things because of its limited profitability and resistance from internal and external stakeholders (such as suppliers, factories, dealers and oil companies).

The next electric car to be put on the market in California (the Tesla) was developed by a new entrant to the field, as electric cars were no longer seen as a strategic priority for incumbents. In 2010, after being saved from bankruptcy by the US government, GM was working again on the development of new models using electricity, such as the Volt.

$\rightarrow$

There are several industry trends that can explain the return to electric technology among the strategic priorities of GM.

First, the general environment faced by the US automobile industry has evolved significantly. Public and financial incentives to buy gas-guzzlers such as the Hummer have been reduced and oil prices have more than doubled. Traffic jams in big cities also create new pressure to regulate vehicle emissions.

Second, US auto manufacturers face increasing competitive pressure from foreign manufacturers, both in their home market and in emerging countries, which are now the main growth markets for autos. The pace at which new models are being introduced is also accelerating, from nearly a decade in the past to less than eighteen months now. Both at home and abroad the business model of the big three (GM, Ford, Chrysler), essentially based on lending consumers money to enable them to buy expensive but traditional large cars and pick-ups, is fast becoming obsolescent.

Third, certainly abroad, but even in the USA, consumers are increasingly sensitive to the environmental impact of the automobiles they buy as well as to their expected running costs. Gas mileage and CO_2 emissions are slowly displacing horsepower and size as the key buying factors for many consumers.

Fourth, new battery technologies, new materials to make cars lighter and even new business models regarding how mobility is purchased (such as models based on total cost of ownership, environmental incentives, car sharing or the leasing of batteries) are threatening traditional approaches to the design, manufacturing and marketing of automobiles.

In other words, global and industry-specific trends are pushing radical innovations back to the corporate agenda, even for a firm (GM) which led the industry for decades and which for years had one of the biggest R&D budgets worldwide.

Hence today, more than ever, managing innovation is imperative to be a growing, profitable company. But as the dot.com bubble beginning in the late 1990s showed, just following a trend without understanding its driver is a source of disasters. Hence we shall discuss here the main trends that put innovation on the chief executive officer's (CEO's) agenda. Those include global trends, which are felt throughout the economy, and industry trends, which are more or less important from one sector to another.

Global trends

The main global trends we can identify are, on the one hand, the growing pervasiveness of innovations, and the new world they announce, and on the other, the growing intensity of international competition.

Global trend 1: innovation becomes business as usual

From pottery to plough, from irrigation techniques to mathematics and writing, innovations that changed entire civilizations have always emerged alongside the evolution of human beings. What has changed, however, in recent decades is the pace and breadth of innovation.

In the past, radical innovations (such as the switch from steam to electricity, or the emergence of the computing industry) might happen only once in the lifetime of a manager and could be dealt with punctual measures. But innovations are now part of the daily life of many firms. No longer an extraordinary event requiring special measures, innovation has become business as usual for managers.

As a consequence, while it was still confined to the R&D laboratories of large firms or public organizations in the 1970s, innovation is now one of the main drivers of productivity growth across industries (before labor and capital investments) and is strongly correlated with wealth at the national level (measured as per capita gross domestic product (GDP)).

According to a recent OECD study,[1] most large firms and between 20 percent and 60 percent of SMEs across developed countries have introduced innovation in the past two years – a significant product or process innovation, or a significant marketing or organizational innovation. As a consequence, increasing numbers of firms must now compete, and more and more managers must now take decisions in what Eric Beinhocker called as long ago as 1997, a "new world".[2]

From the principles of scientific management developed by Frederick Taylor in 1911 to the famous five forces framework developed by Michael Porter at Harvard in 1979, management thinking has long been based on an assumption of an environment where rational expectations and stable equilibria were the rule. Under such a paradigm, successful management was based on the systematic collection of hard facts and figures, within relatively closed and well-defined industrial organization structures. In such a stable world, microeconomic calculations and strategic focus must lead to sustainable competitive advantages, achieved mainly through economies of scale.

The CEO of a firm in this world was like the medieval king, defending a known territory (his market) building big fortresses (his strategy) with thick walls (entry barriers and sustainable competitive advantages).

But, to quote Eric Beinhocker: "When you talk to business leaders today, the notion that any competitive advantage is sustainable has gone away. Success is now based on creating a flow of temporary advantages, and to do that you have to be able to innovate."[3]

Managing in such a "new world" is now about dealing with constantly changing people and ideas, assessing how a complex web of relationships co-evolve through new cognitive behaviors and experimentations. Successful management now requires real-world tests and simulations as well as strategic robustness, allowing firms to achieve continuous adaptation and build flexibility.

In this "new world", yesterday's supplier is the partner and customer of today and might become tomorrow's most aggressive competitor. The CEO of a firm in this world is like the adventurous explorer or pioneer, who travels light and quickly because he or she knows that unexpected enemies or storms can appear.

Of course, hard facts and figures, strategic focus and sustainable competitive advantage have not completely disappeared. But for increasing numbers of firms across more and more sectors, relying only on them becomes less and less viable.

Global trend 2: Innovation crosses borders

The second major trend affecting firms across sectors is the rapid breakdown of most geographical boundaries, linked to the accelerating spread of innovations.

As an illustration, the watermill was introduced independently in various places throughout history and took more than a thousand years to gain wide acceptance. In contrast an upgrade to your web browser can today be installed globally in less than a thousand *seconds*.

What has changed in the meantime can be summed up in three main factors:[4]

- the deployment of cheaper and easier communication and transport infrastructures across the world, such as the internet and international freight;
- the emergence of global segments of customers with similar needs across cultural and geographical boundaries, such as in fashion and luxury goods; and
- the increasing focus of firms on economies of scale and scope in order to drive down costs.

It must be stressed that history and politics have taught us that these three factors cannot be taken for granted everywhere all the time. However, they have consistently led to the international expansion of corporations (for example, in the auto industry), to a relative decline in international variety in many industries (for example, in consumer electronics) and to an increase in international interconnectedness (for example, in the energy sector). As foreign direct investment (FDI) and international trade have grown, so has international competition.

As a consequence, markets and industries such as steel or energy which used to be protected by natural or institutional barriers, and could therefore afford to evolve at their own pace, are now exposed to international competition. In particular, the fate of a firm can now be affected by an innovation that is launched at the other end of the world.

The resulting challenges include:[5]

- shifts of economic activity between and within regions, and the emergence of new customer segments and new industry structures (for example, in the financial sector);
- the emergence of a global market for information and talent as well as the professionalization of management practices (for example, in the entertainment and software industries);
- increased pressure on natural resources, leading sometimes to conflict (for example, in the energy sector); and
- increased spread of communications and exposure to a social backlash (for example, in the oil industry).

These pressures and challenges pose a particular threat to high cost/high value socio-economic systems such those prevailing in Western Europe, where explicit or implicit social contracts link high tax rates with extensive public services, social welfare and retirement systems that are vulnerable to outside threats.

Conversely, it represents significant opportunities for some emerging countries, where entrepreneurial activity is a major avenue enabling people to escape poverty. As a consequence, the infamous BRIC countries (Brazil, Russia – to a lesser extent, India and China) are now not only sources of cheap labor but also of innovations and of growing market opportunities.

In summary, strategies that were based on entry barriers, focus and the protection of competitive advantages are now increasingly under pressure. The deregulation and liberalization of entire industries and the increased importance of intangible assets (which travel easily and can be copied) mean that sheer scale and historical assets often matter less than flexibility and innovation.

Direct evidence of this increased instability is the growing turnover among the top global firms. A firm present among the *Fortune 500* (the 500 largest firms globally) today has about a 50 percent chance of losing its position in that ranking in less than ten years. Similarly, the share of firms within the *Fortune 500* from outside Western countries has more than doubled since the year 2000. This is particularly worrying for large European firms, which tend to be on average much older than their US or Asian counterparts.

Industry trends

Obviously, barriers to entry and industry turnover do not have the same importance across all sectors. As well as the two global trends highlighted above (innovations happen more often and in more places), there are also specific trends that affect firms at industry level. We shall discuss the main ones below.

First, industries see their business environment evolve. New regulations and standards force companies to find new ways to serve their customers profitably; for example, in the car industry, where customers are now more sensitive to energy costs and environmental impact. This includes new product regulations, and rules on safety and environment, intellectual property, privacy and customer protection as well as the successive standards adopted by industries such as in the telecommunications and information technology sectors.

Second, industries experience increasing competitive pressures. The deregulation and liberalization of more and more sectors threaten existing businesses and reduce barriers to entry. This is the case, for example, in the telecommunications sector, where many national incumbents have been attacked by new entrants. This can emerge from the evolution of national economic policies, international trade agreements or simply from the development of communication and transport infrastructures.

Third, new demands and expectations from stakeholders – in particular, customers and shareholders – emerge and require new answers. This is the case, for example, in the food and energy sectors, where security of supply and environmental impact matter increasingly. This can derive from the aging, urbanization or changing social structure of the customer population or from new cultural, national, religious or environmental values and priorities of customers or shareholders. It can also emerge from firms demanding guarantees about the integration, reliability, ecological impact or ethics of their supply chain.

Fourth, disruptive technologies shake more and more industries. Such technologies do not just outperform existing approaches; they can make existing businesses obsolescent. This is the case, for example, in the media sector, where many paper-based or broadcast-based businesses are threatened. Such technologies offer completely new ways to create value, meet needs and complete specific jobs. They often do not come from direct competitors having better R&D or engineering capabilities, but rather by outsiders reinventing the rules that govern industries. For example, Sony led the industry with its Walkman but was then displaced by a firm (Apple) that was initially not perceived as a direct competitor.

Finally, emerging technologies as well as scientific and technological progress and discoveries in areas such as electronics, energy, health care or new materials provide new opportunities for value creation in their own sectors but also in all the industries using their products and services (such as transportation, the media or financial services).

Let us stress that, among these trends, only the last one is dealt with when managers or policy-makers focus only on R&D spending as a measure of the innovativeness of a firm or an economy. We shall revisit this issue in Chapter 4.

The emergence of these trends does not mean that CEOs should place all their energy and attention on innovation alone. What it means is that they should all understand where on their agenda it should sit. What matters for a given firm is how much the trends are affecting its business model.

In particular, a firm must consider where it stands regarding two key dimensions implying different styles of leadership.[6] The first dimension is whether the trend is reinforcing the validity of existing approaches or calls for new responses; and the second is the level of uncertainty:

- A new response needed with a low level of uncertainty implies a focus on change management and judgmental decision-making. This is the case, for example, in the news industry, where traditional advertising-based business models must most probably change or disappear.
- A new response needed but with a high level of uncertainty implies a focus on exploration and strategic innovation (see Chapter 3); it is about inspirational decision-making. This is the case, for example, in the auto industry, where how exactly the automobile of 2020 will look is highly uncertain, but we are pretty sure that it will be quite different from that of 2000.
- An old response which remains appropriate with a low level of uncertainty calls for a focus on effective exploitation and stable growth, or programmed decision-making. This is the case, for example, in the banking sector, where back-to-basics approaches are needed after recent financial crises.
- An old response that seems still to be appropriate but with a high level of uncertainty calls for market intelligence and negotiated decision-making. This is the case, for example, in the construction industry, where it is not clear how long current techniques and approaches will remain valid in the face of increasing urbanization and environmental regulations.

The implication for managers is not that they should all rush to innovate in all directions, but that they should all be aware of the potential impact of global and industry trends on their firm, and of the style of leadership it calls for.

This is particularly the case for traditional large employers in many European countries (in sectors such as transportation, communications, finance, energy, retail, steel or automobiles) which face a dire future if they do not succeed in taking their future in their hands and proactively managing innovations. This means building the right capabilities.

Building the right capabilities

As discussed above, the challenges for managers now is to act less like medieval kings and more like explorers. Of course, scale and efficiency remain important factors for an organization. But competing in the "new world" requires organization to build new capabilities in order to manage innovation effectively.

The first challenge for an organization that wants to manage innovation effectively is to act more like an entrepreneur. This means to be able, within the organization, to identify, assess and deal with new opportunities (and threats). Indeed, the core process of an entrepreneur is to "scan, select and implement",[7] to mobilize resources in order to achieve "leaps forward in the face of uncertainty".[8] This corporate entrepreneurial process must become the engine of innovation within the firm.

The second challenge is to nurture that entrepreneurial process. Whether individual people act like entrepreneurs and form innovative teams, and whether the organization can attract, develop, motivate and retain those people depends on organizational and environmental factors that need to be understood and managed. Those corporate resources must fuel the innovation engine.

The third challenge is to guide the entrepreneurial process. Innovation should not be an objective per se. Not all available entrepreneurial opportunities should be pursued. Innovation should be a means to achieve the strategic objectives, mission and vision that the firm has defined. This means understanding the place of innovation within the firm's strategy, defining how innovative it should be, and what types of innovations it should pursue. This also means sharing this vision across the organization. This shared vision must become the steering wheel of the innovation engine.

From these three challenges we can derive five core organizational capabilities that firms wanting to manage innovations should develop and maintain. These are summarized below (see Figure 2.1) and will be discussed in detail in Part II:

(i) Developing a shared strategic vision of why an organization wants to innovate and which types of innovations it should pursue.

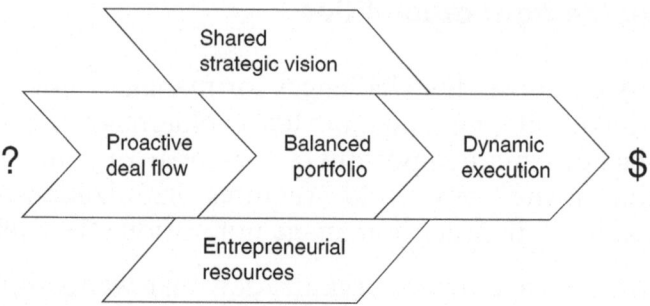

Figure 2.1   Innovation capabilities

(ii)   Attracting and retaining entrepreneurial resources, because innovation is about people and teams, not systems and processes alone.

(iii) Ensuring that a continuous flow of opportunities and threats is identified proactively and systematically.

(iv)   Maintaining a balanced portfolio of projects, based on the assessment and prioritization of the opportunities and threats identified.

(v)   Implementing a dynamic approach to capture the opportunities and deal with the threats in an effective but flexible way.

How these capabilities should be developed and which resources should be mobilized will obviously change from one firm to another and from one industry to another. But for all firms, from the service SME to the industrial giant, these capabilities will define how well they will compete in an innovation-intensive environment.

Key points to take away from Chapter 2

Innovation is now a core element of the competitive environment faced by firms. It is no longer an exceptional "once in a (manager's) lifetime" event, nor the exclusive feature of large industrial firms and their R&D departments. As a consequence, the notion of sustainable competitive advantage built on a careful a priori analysis of a stable environment is fast becoming obsolete. For many firms, strategic robustness and flexibility become more critical than strategic focus and scale.

The natural and institutional barriers that allowed many firms to evolve at their own pace are vanishing. Such firms must now be ready to face competition from innovative companies across the world, or risk being wiped out like the dinosaurs.

A changing business environment, increasing competitive pressures, new needs and expectations, and disruptive and new technologies all affect firms at the industry level. These industry-specific trends amplify the two global trends outlined above and call for a rethink of what is the best style of leadership in each case.

As a consequence, firms should develop innovation management capabilities within their organization. These should be developed around a shared strategic vision, entrepreneurial resources, a proactive flow of opportunities and threats, a balanced portfolio and nimble execution.

Further reading

Becker, W. M. and Freeman, V. M. (2006), "Going from Global Trends to Corporate Strategy", *McKinsey Quarterly,* Vol. 2006/3, pp. 17–27.

Beinhocker, E. D. (1997), "Strategy at the Edge Of Chaos", *McKinsey Quarterly*, Vol. 1997/1, pp. 24–39.

Berkun, S. (2007), *The Myths of Innovation*, Sebastopol, CA, O'Reilly Media, pp. 135–148.

Christensen, C. and Bower, J. (1996), "Customer Power, Strategic Investment and the Failure of Leading Firms", *Strategic Management Journal*, Vol. 17, pp. 197–218.

Davila, T., Epstein, M. J. and Shelton, R. (2006), *Making Innovation Work*, Philadelphia, PA, Wharton School Publishing, pp. 75–79.

De Wit, B. and Meyer, R. (2010), *Strategy: Process, Content, Context*, Florence, KY, Cengage Learning EMEA, Ch. 10.

Foster, R. and Kaplan, S. (2001), *Creative Destruction: Why Companies That Are Built to Last Underperform the Market – and How to Successfully Transform Them*, Strawberry Hills, NSW, Australia: Currency Press.

Hamel, G. and Breen, B. (2009), *The Future of Management*, Boston, MA, Harvard Business School Press.

Lawson, B. and Samson, D. (2000), "Developing Innovation Capability in Organizations", *International Journal of Innovation Management,* Vol. 5, pp. 377–400.

Shane, S. and Venkataraman, S. (2000), "The Promise of Entrepreneurship as a Field of Research", *Academy of Management Review*, Vol. 25, pp. 217–226.

Smith, D. (2006), *Exploring Innovation*, Boston, MA, McGraw-Hill, pp. 252–269.

Tidd, J., Bessant, J. and Pavitt, K. (2005), *Managing Innovation*, 3rd edn, Hoboken, NJ, John Wiley, pp. 89–97.

Synthesis of Part I

This first part focused on understanding what innovation means from a business point of view and why it matters. We highlighted that innovation means much more than creativity or inventiveness, and that newness, from an innovation management point of view, is relative.

We defined innovation as having two facets: newness and change. We discussed how the importance of the change dimension of innovation means that adoption and diffusion are key processes that need to be managed if one wants to disrupt competitors rather than customers. As disruptive innovators and the incumbents they threaten have discovered, what matters most is who can make change happen, not who can come up with the most sophisticated inventions.

We highlighted that managing innovation means much more than developing revolutionary products. Innovation is about what you offer (products, services and business models) as well as about how you deliver it (value chain, organization and partnerships, strategy).

Innovations will also differ according to the level of change they generate, from incremental exploitation of business-as-usual activities to the revolutionary exploration of new business arenas. In particular, what matters in terms of the management of an innovation is the level of disruption created by the change it generates, not the inventiveness or sophistication of the underlying technology.

Innovation must be on the agenda of all firms because increasingly it is affecting the economic development of all sectors across all regions. Managers must understand how those trends as well as the evolutions of their industry affect the importance of innovation and the success of their business. They should accordingly develop the right innovation capabilities, building, nurturing and guiding entrepreneurial processes.

In Part II we shall detail what those innovation capabilities actually entail and what are the tools and frameworks that can be used to develop them in a systematic and pragmatic way.

PART II

DEVELOPING INNOVATION CAPABILITIES

A very common mistake for firms focusing on innovation is to see it as an objective *per se*. Accordingly, such firms put in place processes and systems to boost innovation and, implicitly or explicitly, define the quantity of innovations (such as the number of ideas, the number of patents or the number of projects) as a key performance metric of the organization.

But what this often generates is costly compliance behavior (people generate ideas, patents or projects because they are asked to do so, not because it makes sense to them), a dispersion of resources across numerous more-or-less useful projects (lack of focus) and a loss of momentum and motivation, as projects fail to improve business performance.

What should matter first is why a firm wants to innovate (strategy); who is best positioned to make it happen (entrepreneurial resources); and what are the systems and processes that can be put in place to identify, assess and implement effectively innovation opportunities (proactive flow, balanced portfolio and nimble execution). These five capabilities will be discussed in the five chapters of Part II.

The first innovation management capability to consider in a firm is therefore whether it can define and share a clear strategy where innovation is a means (integrated with other means) of achieving defined objectives, both at corporate and business level.

CHAPTER 3

Strategy: how and how much to innovate

Short case: strategic innovations in the airline industry

The airline industry was for decades a relatively stable industry, based on national or regional incumbent players. These companies more or less all competed to offer the best connections and the best in-flight experience, in particular to high-margin (less price-sensitive) business customers. Hence most airlines used a similar strategy, based on vertical integration and multiple connections offered through their base airport (the hub). The industry as a whole was actually not profitable most of the time. National regulations, state support to national champions, strong unions and a lock on the best landing slots (in particular, for the most profitable transatlantic flights) all reinforced the status quo.

Then the partial deregulation and globalization of air travel, the emergence of entrepreneurial companies and of new regional competitors such as high-speed trains, combined with the decrease in state support, all contributed to major shake-ups of the industry. Stuck with complex and unprofitable business models, many airlines tried to develop innovative strategies.

From a corporate strategy point of view, this led on the one hand to the consolidation of the industry through various

→

mergers, acquisitions or partnerships developed in order to generate synergies and global offers. On the other hand, this led to a vertical fragmentation of the industry, with some players developing new businesses focusing on specialized services such as freight, aircraft maintenance, reservation systems, handling or catering.

From a business point of view, this led to a diversification of value propositions aimed at various segments (such as leisure travellers, commuters, one-way tickets or last-minute bookings) and a complete reengineering of the value chains. Hence you can today buy a ticket from one airline, travel on a plane leased by another, be served by flight attendants from a third airline, and be seated next to a person having paid a very different price for the same flight. This shake-up also led to the emergence of low-cost players based on radical business strategies and entirely different business models. Some of these new players are now among the most profitable airline companies.

Hence the strategic role of innovation can be defined at both the corporate and business level. In this chapter we shall first outline the strategic value creation levers where innovation could play a role (see Figure 3.1) and then discuss the factors that will influence which strategic levers should be used and how innovative the strategy should be.

The key issues at the *corporate* strategy level are about defining, on the one hand, the scope or perimeter of the firm, and on the other how to make the parts worth more than the whole. In other words, this is about defining where (in which businesses) the firm wants to create value.

The key issue at the *business* strategy level is defining for each business the core elements of the business model. These include, on the one hand, the product/market positioning (who are the customers, what is the value offered to them, and when will the business compete?), and on the other hand the management

Figure 3.1   Strategic value creation levers

of the value chain (how will the business deliver value to the customers?). We shall discuss below the role innovation can play in each of these issues.

How to develop innovative corporate strategies

The first question that firms must address at corporate level is in which businesses they want to be actively involved. This means deciding whether (and when) to stick to the current core business (*exploit*) or renew it (*explore*).

Exploitation **versus** *exploration* **corporate strategies**

The first option (*exploitation*) can be considered to be adapting to the dynamics of the industry and trying to play by its rules. It is about the refinement of existing solutions in order to maintain and improve performance. Such an approach leads the firm to focus on its "core business"[1] – that is, the collection of resources, capabilities and processes which provides it with its unique strength over its competitors. This strength is achieved in particular by leveraging its collective routines, know-how and culture developed through substantial experience and learning.

Firms focusing on such exploitation strategies will tend to promote innovative projects aimed at developing their long-term competence platforms, defend existing businesses in a focused and systematic way, and try to catch low-hanging fruits. They tend in general to be both prudent (low risk) and impatient (quick results).

Scholars and consultants have defended such an approach based on the fact that firms will create value only in the businesses where they can claim a unique competitive advantage, otherwise they will lose focus. As illustrations, firms such as Exxon, Ferrero (the producer of Nutella) and Coca-Cola have for decades created value by sticking to what they know best.

The second option (*exploration*) can by contrast be considered as a way to achieve industry leadership and attempt to set new rules (shaping rather than adapting). It is about experimenting with new ways of improving organizational performance. It means managing the creative destruction process proactively rather than trying to slow it down.

Firms focusing on exploration strategies would rather develop innovative projects leveraging the competences of partners, be ready to cannibalize their existing businesses and make bold moves. They tend to be both radical (high risk) and persistent (long-term view). Such firms will constantly try to find adjacent growth opportunities – for example, through new value chains, new products, new geographies, new channels, new market segments or in some cases entirely new types of business.

Scholars and consultants have defended such an approach based on the observation that, over the long term, competitive forces prevent firms from maintaining their value creation potential. Indeed, across industries new entrants tend to be more profitable than incumbents. Such an analysis sees commoditization as the end game for all industries. There is therefore a constant need to create new competitive spaces and escape "competency traps".[2] As illustrations, firms such as Nokia and Apple have created substantial value by redefining the nature of their core business several times.

In summary, exploitation can be compared to a mining process, leading a firm to try to extract all the opportunities available in its current position. Conversely, exploration can also be compared to a hunting process, leading an organization to look outside its core business (its territory) for new ways to answer new market demands.

Two conclusions can be drawn from this debate. First, the two alternatives presented above (exploration and exploitation) represent extreme archetypes. As their environment and resources evolve, firms will adopt different compromises between the two extremes, going through successive "hunting" and "mining" phases. Exploitation cannot be pursued for ever, as new opportunities provide decreasing returns, but focusing on exploration alone would often be too risky and costly in terms of resources. Managers must therefore continuously balance the two extremes and define how innovative their corporate strategy should be.

Second, while exploration strategies provide new opportunities to grow, it would be a mistake to see them as substitutes for exploitation approaches. Even as firms try to expand or diversify, they should maintain sufficient "mainstream" capabilities (see Chapter 4) to defend their home market, generate healthy organic revenue growth and achieve improvement opportunities. In other words, one can build new floors only if one maintains strong foundations; one can develop star businesses only if one can rely on some cash cows. The temptation, however, is often high (especially in small entrepreneurial firms) to overinvest resources and management attention in new developments at the expense of the viability of the core business. Here, again, reaching the right balance is what matters.

Having considered the businesses in which they want to operate, the second question firms must address at corporate level is what should be the shape of the business portfolio.

Shaping the business portfolio

Once the set of existing and new businesses has been defined, the next key challenge is to make sure that the whole is worth more than the sum of the parts. This can be achieved through a balanced allocation of resources across the portfolio of businesses, by the development of shared corporate resources, and through innovative portfolio design. We shall discuss those three aspects below.

First, as outlined above, the balance of financial but also managerial resources must be monitored carefully across the portfolio of businesses, in particular balancing exploration and exploitation businesses. Various frameworks, such as the BCG matrix[3], or the GE–McKinsey matrix[4] have been developed to support such decision-making processes.

Second, managers must consider how strategic assets can be developed and shared at corporate level in order to support the businesses, and in particular the way those businesses manage innovation. Those strategic assets include:[5]

- Business platforms providing access to adjacent businesses, including noncore businesses and orphan products, such as support organizations or supplier networks. For example, utilities across the world are exploring how the large infrastructures they have built over time can be better exploited.
- Untapped customer insights, such as privileged relationships and distribution networks or unique customer data and information that can be exploited across business segments. For example, telecom firms and financial institutions try to improve cross-selling and expand their range of products or services by better leveraging their customer base.
- Underexploited capabilities, such as proprietary technologies, standards, functional expertise or infrastructure, or market and competitive intelligence. For example, many retailers attempt to exploit their distribution networks in novel ways, developing new offers in telecommunications, banking or even automobiles.

Finally, a firm can in some cases create value through an innovative business portfolio structure. While the business portfolios of competing firms are often of a similar shape, managers should consider developing alternative approaches. Innovative business portfolios involve, for example, remaining fragmented, close to market businesses despite the industry becoming concentrated, developing international or horizontally integrated portfolios while most other companies are integrating vertically, or developing strong links with businesses from other industries.

As an example, Vodaphone implemented in the 1990s a corporate strategy based on the horizontal integration of cell phone businesses across several countries. This innovative business portfolio structure initially provided the firm with extensive economies of both scale and scope, in particular in terms of bargaining power *vis-à-vis* handsets and network infrastructure providers. However, as the industry evolved towards the integration of cell phone offers with broadband internet access and interactive TV services, this phone-only corporate structure became a liability, as Vodaphone struggled to compete with the bundled packages offered by local vertically integrated players.

In summary, managers can be innovative at corporate strategy level by:

• implementing new trade-offs between exploration and exploitation – for example, through aggressive new business development activities or, on the contrary, by specializing in a narrow field;
• finding new ways to develop and leverage corporate resources across their businesses, in particular through shared value chain or shared product/market positioning; and
• designing a portfolio of business with a novel structure or shape, uncovering new sources of synergies and/or economies of scope or scale.

Let us stress, however, that innovative corporate strategies are usually much easier to design (newness) than to implement (change). Indeed, managers often struggle to deal with core rigidities and to devise a workable definition of the strategic business units and of their interdependencies. Such core rigidities include underlying business conditions hindering integration or restructuring, and resistance from the existing power structure, risk aversion or inertia. Many firms, in particular in the utilities (diversification) and airline industry (emulating low-cost business models) have indeed failed to implement innovative corporate strategy despite those strategies being in theory quite attractive.

Having reviewed the key levers for innovative corporate strategies, we shall discuss below how innovative business strategies can be.

How to develop innovative business strategies

The first question firms must address at the business level is to define what they want to sell and to whom (product/market position). This means deciding whether the firm will continue to sell similar offers to the same market or to develop new offer, enter new markets, or both.

The second question firms must address at the business level is to define how a business paces the development of its product/ market position. It means deciding in particular whether the business wants to be a first mover or a follower.

The third question firms must address at the business level is to define how it wants to deliver value (value chain). This means deciding in particular whether (and how) it should improve the way it designs its product, delivers it or deals with customers, or whether it should introduce completely new ways to run the business.

We shall discuss these three aspects below, plus their integration with innovative business strategies.

Innovative product/market positioning

The first key aspect of a business strategy is the definition of what is offered and to whom; in other words, what are the value propositions developed by the business, and which are the target markets. From an innovation management point of view, businesses must therefore consider whether they want to develop innovative product/market positioning.

As an illustration, a financial firm offering payment and intermediation services can consider whether to improve them or to move into new insurance or advice activities. Similarly, if a financial firm is developing offers to both individual and corporate clients, it can consider strengthening its position in these segments or moving into new private banking, public-sector financing or niche segments such as SMEs or expatriates.

Hence a product/market positioning can be innovative in terms of new offers and new markets. Businesses can try to deliver new or different benefits by introducing either variations of the existing offer or completely new ones. These can be developed by modifying the core product or basic offer, but also by modifying what customers perceived as the augmented or expected offer.

This is the case, for example, for drinks companies who want to increase their "share of throat" among their customers (for example, a beer company offering fruit juices or a cola company offering water), or for automobile companies such as Renault or BMW trying to sell completely new driving experiences rather than just simply cars.

Alternatively, businesses can try to leverage in-house capabilities and replicate success by serving new markets, such as the unmet or unsatisfied needs of customer groups. This can be done by increasing the businesses' share of available customers or by acquiring new ones. This can also be done by targeting new segments within the same market, or by internationalizing domestic businesses. This is the case, for example, for food companies targeting ethnic or religious groups, or for travel agencies targeting specific age groups.

Let us stress, however, that many businesses choose to stick to their current product/market position and try to improve it; for example, by streamlining existing offers and activities or bench-marking best practices. This is the case, for example, in stable, mature markets such as those for raw materials, where firms try to consolidate and improve their position, or in local markets, where domestic firms replicate existing business models.

According to the OECD,[6] about 5 percent of innovative firms are "adopters", who replicate existing business models, and 9 percent and 11 percent, respectively, are domestic and international "modifiers" who introduce variations in their home market or abroad. Another 5 percent and 12 percent respectively of (generally larger) firms are domestic and international "new-to-market innovators", who develop new offers at home or abroad. Finally, between 35 percent and 80 percent (depending on which OECD country is being considered) of the businesses surveyed were not considered to be innovative – a surprising total.

Once the business strategy in terms of product/market position has been defined, managers must decide about the timing or pace of their strategic moves. This will be discussed in the next section.

Timing of business innovations: the hare and the tortoise

A common strategic mistake made by businesses is to assume that, from an innovation point of view, first is always the best. Popular beliefs such as "time is money" and "winner takes all" make them assume that there is always an intrinsic benefit linked with being first in a market.

But decades of empirical evidence across industries have demonstrated that this is not always the case. While testing customer response as early as possible or decreasing development costs have obvious benefits, being the first to enter a market has both benefits *and* drawbacks. Indeed, in many industries the "second players" have made more profit than the industry leaders.

The advantages of being a first mover relate to the fact that once an offer is developed, it can be difficult for competing businesses to catch up or displace it. First movers can enjoy brand loyalty and technological leadership by building up R&D investments and intellectual property – such as in the auto industry.

First movers can also preempt markets and scarce assets (locations, natural resources, key suppliers or distribution channels) by capturing the best positions. This is the case, for example, in the satellite business, where the company that captures the best orbital "slot" will be the most profitable.

First movers can create high switching costs for customers – for example, because of high learning and adaptation costs, exclusivity contracts or simply routine. This is the case in the retail banking sector, where customers often stick to the first bank they dealt with (sometimes as teenagers), the bank with the closest branch or the bank their parents use.

Finally, first movers can in some cases reap increasing returns advantages by capturing dominant designs or leveraging network externalities, such as in the software (for example, for operating systems or business suites) or e-commerce sectors (for example, for online auctions or social networks).

As an illustration of these first mover advantages, leading search engine providers can capture significant first mover advantages as they build unique technological skills and brand recognition, develop unique computing infrastructure and partnerships, generate significant user switching costs (through bookmarks, home page or simply routine) and finally leverage network effects in advertising.

However, first movers can be less profitable than "fast followers", because those followers might avoid some of the investments and risks that first movers have to incur. Moreover, enjoying first mover leadership in one market can create strategic commitments that are difficult to unfold when that market evolves.

Indeed, followers can benefit from information leakages (in particular through the poaching of people) and free ride the investment first movers had to make in terms of R&D, infrastructure development or buyer education.

This is the case, for example, regarding the development of disposable razors, where once the concept was tested and proved many tried to copy it.

Followers also benefit from the fact that first movers often have to resolve uncertainties and explore alternative options. Once these uncertainties are solved, a fast follower knows what to do in terms of technological feasibility or customer requirements. This is the case, for example, in buildings or aircraft manufacturing, where experimental designs can be copied if they prove to be successful.

Moreover, first movers will often build leadership positions, in particular in terms of market share. However those positions can decrease their ability to react to future market evolutions, because they will have the most to lose.

This inertia or "lock-in" effect can be linked to fears of cannibalization. This is the case, for example, when leading compact disc or tape music player manufacturers failed to develop digital music players. It can also be linked to inflexible commitments to the supply and distribution channels, manufacturing infrastructure and complements, or simply routines. This is the case, for example, when leading retailers and consumer goods businesses fail to embrace e-commerce opportunities.

As an illustration, auto manufacturers often fail to benefit from significant first-mover advantages, as competitors can free ride their research and investments through information leakage and reverse engineering to resolve uncertainties. For example, when Renault successfully introduced a European van (the Renault Espace) or when BMW created the small-luxury segment (with the Mini) while Mercedes struggled with its Smart range, others quickly followed and developed similar offers. Furthermore, automobile manufacturers often face significant first-mover disadvantages, as their leading position in some markets generate fears of cannibalization as well as commitment to specific designs, supply chains, unions, and distribution and customer service channels (see the auto industry short case in Chapter 2).

As a consequence, businesses should understand that, when managing an innovation, first is not always best. They should in each situation assess the benefits and drawbacks of being a first mover, a fast follower or a late entrant before investing resources to try to be a pioneer. Finally, once the timing of entry is defined, they should identify and leverage the benefits linked with that timing while understanding and mitigating the drawbacks.

In the case of a businesses intending to act as a first mover, this can be done or example by enforcing secrecy and patent protection, for example, investing in learning curves and building up complementary assets, lead-time and after sales services. This maximizes some of the advantages of being the first mover while mitigating the corresponding disadvantages.

Having defined what to sell to whom and when to do it, businesses should consider how to be innovative regarding how to deliver; that is, how they manage their value chain. This will be discussed in the next section.

Innovative value chains

Whatever their size or sector, all firms have to engage in four types of activities to deliver some added value to their customers (see Figure 3.2):

- Designing the value proposition; that is, defining how the product/market position of the firm evolves and at what rate (see above). This might include, for example, R&D and strategic marketing activities.
- Delivering the value proposition and managing operations; that is, realizing all the steps necessary to make the value proposition available to customers. This includes in particular the procurement, transformation and logistics of the goods and/or services on offer.
- Dealing with customers, both existing and prospective. This includes sales, operational marketing and customer service activities.
- Dealing with the organization itself and its partners through support functions. This includes infrastructure and systems management, human resources and finance management, public relations and legal activities.

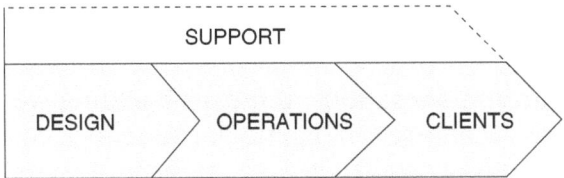

Figure 3.2 Value chain

As an illustration, an airline company must manage its network of destinations and its flight schedule (the value proposition), its aircrafts, crews and passengers services (the operations), its ticket sales, customer complaints and advertising (the customer management) and finally its computing systems and human resources (the support).

Of course, the relative importance and complexity of these four types of activities and whether they are formally and explicitly organized (whether they are in somebody's job profile) will vary from one firm to another and from one industry to another. They are, however, always present.

From a business strategy point of view, managing innovation in the value chain should therefore mean deciding how each of those activities should be performed, and in particular whether they should be performed within the business itself or out-sourced (make or buy).

It means deciding whether the way that each of the activities is currently performed should be improved or completely renewed, and what are the key performance metrics to consider.

Regarding the design activities, this means in particular deciding whether new processes could be put in place that will improve the perceived quality or the adaptability of the value proposition.

The clothing industry offers an example of this – it has improved its ability to keep up with the evolution of fashion by tracking closely the activities of its distribution channels and developing flexible partnerships with its suppliers. The biotech industry too has developed new ways of designing drugs by using genetic profiles and rapid throughput screening techniques rather than the traditional hypothesis-driven experimental design that was used previously in most laboratories.

Second, regarding operations, it means deciding what could be improved or restructured, in particular to reduce costs, increase quality or react more swiftly to variations in demands.

The steel industry is, for example, reconsidering its entire manufacturing and logistics process to better manage energy efficiency and resource savings as well as implement safer, cleaner and less capital-intensive processes. In yet another example, the wine and beer industries have seen new manufacturing processes competing with traditional ones.

Third, from the clients' point of view it means deciding whether the way the business interacts with existing or potential customers could be upgraded or revolutionized, in particular in order to improve trust and the quality of existing relationships.

The internet has led some automobile firms to redefine the role of automobile dealerships completely, as many potential customers now first go online and know a good deal about the auto they want to buy before they physically enter a dealership. Similarly, many banks have redesigned their branches, with simple transactions being migrated to direct channels such as ATM or online banking services.

Finally, regarding support functions, it means deciding whether the cost and quality of those functions could be improved. For example, investments in information technology have revolutionized the way payments and orders are handled and accounted for, sometimes by completely externalizing those functions. Similarly, human resource management has morphed in many organizations from simple payroll administration to strategic talent management.

The strategic mistake many businesses make here is again seeing innovation too narrowly, by focusing, say, only on new manufacturing or packaging processes. While such types of engineering processes attract the most attention, there is often plenty of untapped value to be found through innovative partnerships, innovative sourcing or logistics processes, complementary services, new routes to market, or brand and communication innovations.

We have considered here how businesses can compete by improving their product/market positioning or their value chain through innovation. We shall now consider more radically innovative business strategies, where the entire business model (both the product/market position and the value chain) is redefined.

Business model or "strategic innovation"

Businesses can develop temporary competitive advantages by developing an innovative product/market positioning (for example, in the luxury goods industry) or innovative value chain (for example, in retail). The concept of business model or strategic innovation aims at radically changing both. Its aim is to create new competitive spaces rather than to fight for existing ones.

Strategic innovation can be defined as "the capacity [for a business] to reconceptualize existing industry models in a way that creates new value for customers, get ahead of competitors and produce new wealth for all stakeholders".[7] It involves a "fundamental re-conceptualization of what the business is all about, which, in turn, leads to a dramatically different way of playing the game in an existing business".[8]

Rather than benchmarking their performance and trying to catch up or overtake competitors, strategic innovators "make the competition *irrelevant* by offering fundamentally new and superior buyer value in existing markets and by enabling a quantum leap in buyer value to create new markets".[9]

Traditional business strategies tend to take the competitive environment for granted and focus on finding the best compromise between differentiation and cost leadership postures. Such businesses try to manage the (given) trade-offs effectively between, on the one hand, enhanced value and features, precision and quality, target niche markets and economies of scope, and on the other minimum features, logistics and manufacturing efficiency, and economies of scale.

In a strategic innovation approach, firms consider the industry environment (political, economic, socio-cultural or technological) and the resulting trade-offs between differentiation and cost leadership not as given but as an endogenous factor. A strategic innovation approach is about shaping new contexts of consumption and competition (changing the rules) rather than adapting, about developing industry leadership rather than following industry dynamics.

As an illustration, the automobile industry was long defined by most operators developing a similar range of offers, going from the small and cheap to the big or fast and expensive. Each new model was therefore benchmarked and competing directly with existing competitors' models offering similar cost/quality trade-offs (such as the BMW Series 3 with the Audi A4, the Volvo S and the Mercedes C-Class). The industry itself was analyzed by most operators along rather fixed market segments (the family auto, the urban auto, and so on). Then new models such as the Renault Espace, the BMW Mini, the Mercedes Smart and the Toyota Prius Hybrid were launched. These were not just improvements on existing model ranges and did not fit into any of the industry's historic segments. As a consequence, they did not initially face competition from similar models. Indeed, they succeeded in changing the criteria used to market an automobile, to create new market segments and to redefine how a vehicle can be perceived as a value proposition.

While opening new, uncontested space looks very promising, the actual implementation of a business strategy based on strategic innovations can involve significant risks and challenges.[10] It means being ready and able to test unproven business models, often focusing on industries with a high revenue growth potential but that are still fuzzy or emerging. Hence such strategy can be unprofitable for several quarters or even longer, with no clear picture of performance in the early stages (but with visible costs!).

Strategic innovation also means using some of the business assets and competencies but in a radical departure from existing business models. It also often means developing completely new knowledge and capabilities to achieve discontinuous rather than incremental value creation. These discontinuities often generate great uncertainties across multiple functions in the value chain.

The importance of the challenges linked to the strategic innovation approach is illustrated by the limited numbers of clear success stories to which its proponents can point. It offers, however, a refreshing and thought-provoking way to define innovative business strategies. It can therefore be considered as a high risk/high potential strategy, suitable in its pure form only for the firms with the right resources and expectations who are facing a particular competitive environment.

Whether the strategy of a business should be highly innovative in such a way will be discussed in the next section.

How innovative should your strategy be?

Having identified the various ways innovations can be integrated into the strategy of a firm, at both corporate and business levels (the levers: see Figure 3.1), the next question is to decide how much it should be integrated – in other words, how innovative should a strategy be (how much to pull the levers).

A common mistake is for some firms to consider that innovation must be the first priority, always and everywhere. In that case innovations become an end *per se*, rather than means of achieving existing strategic objectives. Such firms to some extent repeat the mistakes made during the dot.com bubble, when being online also became an end per se rather than just the means it should be. Similarly, before the financial crisis many firms embraced innovative financial products without a clear understanding of the risks and potential consequences.

Indeed, mixed empirical evidence is available regarding a direct link between innovativeness and firm performance. If innovation appears to be a necessary condition to compete in more and more industries (see Chapter 2), whether being the most innovative always means achieving the best performance is not clear. In other words, it is probably not true that high innovativeness is a sufficient condition for success.

As an illustration, the firm that produced the biggest corporate profit ever (Exxon) achieved its performance by implementing known processes effectively in its traditional businesses. Similarly, some tobacco companies, utilities and professional service firms have proved to be highly profitable businesses while not being particularly innovative (in relative terms).

Indeed, developing innovation management capabilities should clearly be high on the CEO's agenda (see Chapter 2) but investing in innovation projects should not necessarily always be the number one priority. The strategic priority given to innovation initiatives by a firm should be defined as a function of the external environment the firm faces, its internal resources and capabilities, and its expectations in terms of risks and performance. We shall detail these three aspects below.

External analysis: competitive and socio-economic environment

Scholars of strategy have long identified the environment a firm faces as a key driver of its strategy, defining both the opportunities it could pursue and the threats it must deal with. As a direct consequence, the strategic role of innovation should also be driven by the characteristics of the environment the firm faces.

The competitive intensity of the firm's environment and the maturity of its industry will affect in particular how much its sales and margins are under pressure. It will therefore influence how much it should innovate in order to develop new markets and avoid the commoditization of its current offers. The competitive intensity will also affect the economics of the firm's business model and therefore its ability to invest in innovation.

This competitive intensity will be affected by, following the famous five-force framework developed by Michael Porter, (i) the degree of consolidation or fragmentation of its market, (ii) the relative bargaining power of suppliers and (iii) buyers, and finally by the prevalence of (iv) alternatives and (v) new entrants. For example, the producers of agricultural products selling to large retailers, or the tier-one and tier-two suppliers of the automobile industry often experience tight margins that put pressure on their ability to invest in innovation initiatives. Conversely, a firm enjoying monopolistic rents might feel less pressure to innovate. Finally, the appetites of existing customers for innovation or, in contrast, their relative inertia or conservatism, will also influence how much the firm should innovate.

The wider socio-economic environment the firm faces will also influence how and by how much it should innovate. This includes, from a technological point of view, the capabilities and infrastructure available and the development of complementary assets in its ecosystem (see Chapter 4), as well as the rate of technological changes prevailing in its industry.

For example, the way firms innovate in the micro-electronic or the aeronautical sectors, despite both being R&D-intensive are indeed quite different.

The socio-economic environment also depends on policy and economic evolutions that affect the feasibility and attractiveness of innovations. This includes macro-economic factors influencing the underlying economics of innovation projects (such as inflation, interest rates or currency movements) as well as the level of political support or resistance to some types of innovation. An example of this was when European governments supported the development of innovations in the field of mobile telecommunication while rejecting some innovations related to genetically modified organisms.

Finally, social and cultural evolutions regarding what customers (and voters) value, support or fear also have a significant effect on how firms should innovate. This includes macro-evolutions such as migrations, aging of the population or nutrition issues (for example, an increase in obesity), or changes in mood regarding, for example, the protection of the environment or the development of nuclear energy.

In summary, the socio-economic and competitive environment a firm faces influences the strategic choices it can and should consider, and consequently the place that innovation should take within those choices.

As an illustration, a bank in the middle of a financial crisis should not consider innovation in the same way as does a utility facing obsolescence or deregulation, or an SME active in a fast-changing technological environment. While this might seem obvious, many firms still innovate for the sake of it, without considering explicitly how it fits with the environment they face and how changes in the environment interact with changes in their innovation strategies.

But scholars of strategy have highlighted that the strategic performance of a firm is not only driven by its ability to position itself effectively in its environment but also, and in some case mainly, by its ability to develop and maintain key resources. In other words, the external environment matters but the internal resources of the firm itself also matter. This is particularly the case for firms facing fast-evolving environments, where a good position today can be a very dangerous one tomorrow (see Chapter 2).

Hence the second element that drives how innovative a strategy should be the resources and competencies available to the firm and how they are combined in the value chain.

Internal analysis: resources and competencies

The privileged assets, unique competencies and special relationships a firm enjoys will define the strengths it should leverage and the weaknesses it has to deal with, and thus how innovative its strategy should be.

Such resources include, first, the human resources and organizational capabilities of the firm and how entrepreneurial they are (see Chapter 4). These capabilities will depend on available expertise and experience; social capital and networks developed over time; the motivation and energy; the prevailing leadership behaviors; and finally the demographics (cultural mix as well as ages and tenures) of the people on whom the firm relies. In particular, the lack of managerial resources (in terms of skills or full-time equivalent) has long been known to be a key factor constraining the ability of a firm to grow (the "Penrose effect"[11]).

These resources also include the tangible and intangible assets of the firm and how they are able to support the identification, assessment and implementation of innovation opportunities (see Chapters 5 to 7). What matters in particular is how much these resources allow the firms to develop dynamic capabilities, which are the "ability to sense and then seize new opportunities, and to reconfigure and protect knowledge assets, competencies and complementary assets to achieve sustained competitive advantage".[12]

The dynamic capabilities of the firm will depend on the size and quality of the assets and facilities available (in particular, the underutilized ones) and on the cash flow and sources of funding generated by the economics of the current business model. They will also depend on the technology and intellectual capital owned by the firm, its brand, its reputation and its customer base as well as its underlying culture and values. As an example, an incumbent postal services firm facing significant threats from the liberalization and disappearance (through the use of electronic mail and invoicing systems, among others) of its business should attempt to leverage its unique resources. These include its distribution networks and logistics capabilities, its brand recognition and customer base, and the fact that it is often the only organization that every single day is in direct contact with most of the people in a given region.

Having considered how the environment and resources a firm faces can influence its innovation strategy, we will now discuss the importance of purpose and expectations.

Prospective analysis: purpose and expectations

Scholars and policy-makers are often puzzled that many large firms engage in mergers and acquisitions that destroy value. Similarly, many SMEs do not grow though they have plenty of value creation opportunities available to them.

The simple answer is that the strategies firms adopt are influenced by the expectations of their stakeholders, which are often more complex and diverse than merely the creation of short-term financial value. Some large firms want to build an empire or pursue market share leadership for the sake of it, for example, while many SMEs do not grow simply because their owner-manager does not want to. To quote Peter Drucker: "There is [indeed] no "perfect" strategic decision. One always has to pay a price. One always has to balance conflicting objectives, conflicting opinions, and conflicting priorities. The best strategic decision is only an approximation – and a risk."[13] As a consequence, when considering its innovation strategy it is critical to take into account the purpose or objectives the firm is pursuing; in particular:

- whether the focus of the firm is on growth (absolute sales or margins) or on profitability (relative return on equity or on investment);
- what is the level of risk and uncertainty that the various stakeholders are able and ready to deal with; and
- what is the time horizon considered – in particular, the short-term versus long-term focus.

Indeed, the innovation opportunities pursued by an entrepreneurial firm focused on long-term absolute growth should obviously be different from those pursued by a risk-averse firm focused on short-term relative profitability.

In particular, how innovative a strategy should be will be influenced by the discrepancies identified between what the firm wants to achieve and what it will probably achieve.

By confronting their objectives in terms of performance and what is likely to happen in a "business as usual" scenario, firms can explicitly identify and quantify the gap – if any – that innovation opportunities should to close. Future margins can be under pressure because of price erosion and cost inflation, calling for significant innovations in the value chain. Similarly, future sales can be affected by competition and cannibalization, which call for significant product and market innovations. For example, firms like Procter & Gamble (P&G) or 3M regularly quantify and communicate explicitly the billions of dollars of future new sales that will have to be generated through innovations.

Moreover, whether a firm will be able to fill those performance gaps through innovation is strongly linked to the ability of its stakeholders (and in particular, the shareholders) to face the resulting risks. The appetite of investors for risk and notably the ability and willingness of the management to raise or set aside venture capital resources will clearly affect how it will manage innovations.

As an illustration, the ups and downs of financial markets and the resulting evolutions in the mood or optimism/pessimism of investors and managers clearly influence the type of innovation strategy that can and should be pursued.

Innovation strategic postures

Innovation should not be the first priority of all firms, at all times, and everywhere. The diversity of ways a firm can be innovative from a business and a corporate strategy point of view as well as its purpose and the specific resources and environment it faces means that each firm must identify the best strategic posture it can adopt regarding innovation.

Firms like L'Oréal or Coca-Cola have been very profitable through market and/or market share growth driven mainly by continuous and mainly incremental product/market innovations. Others, like Southwest Airlines or Dell, have built their success on the effective implementation of radical value chain innovations. Firms like Telefonica or Nokia have achieved substantial value creation through a radical renewal of their corporate strategies, while others, like Volkswagen or LVMH, have introduced new ways to manage their portfolio of businesses, leveraging common platforms and resources.

Hence there is no unique way to be successful through innovation. A firm aiming to improve its profitability might focus on incremental process innovation throughout its organization, while a firm aiming at revenue growth might invest in incremental product/market innovations across its businesses. Firms willing to create entirely new businesses might dedicate specialized resources to develop radical product/market innovations. Finally, new entrants or laggards wanting to leapfrog their competitors might develop specialized capabilities to implement radical process innovations.

What matters is that these choices fit into a coherent strategy and innovation posture that is shared throughout the organization and adopted by the relevant stakeholders.

A firm adopting an offensive innovation posture will consistently build capabilities to be a first mover. A more opportunistic firm will look for some unique unmet needs in a market segment. A defensive firm will choose to wait for a competitor to introduce a product first and resolve some of the uncertainties, then correct any mistakes made and leverage complementary assets.

As an illustration, Microsoft is famous for being very successful in markets (with Windows and Office), where most innovations were initiated by others (defensive posture), while Nintendo (with the Wii) was comparatively successful even if the products of its competitors delivered higher computing and graphic performance (opportunistic posture).

Similarly, firms adopting imitative or dependent postures will produce a clone or a satellite of the pioneer's product, leveraging lower labor costs, access to raw materials or strong manufacturing capabilities. Finally, traditional firms will make minimal changes to products, striving to offer the lowest cost possible. Many leading Japanese firms were initially developed using such copycat strategies.

Key points to take away from Chapter 3

A firm should not pursue innovation opportunities as an objective *per se* but should understand how managing innovations effectively can contribute to its strategic objectives, at both corporate and business levels.

First, innovative *corporate* strategies include new ways to balance exploration and exploitation activities across the business portfolio, new ways to leverage corporate resources across existing and new businesses, and new ways to structure the business portfolio.

Second, innovative *business* strategies can be implemented through new or adapted product/market positions and a new or adapted value chain. When implementing product/market innovations, first is not always best. Firms should understand, assess and proactively manage the benefits and drawbacks linked with (not) being a first mover in a given market.

Value chain innovations can be implemented through new or adapted ways to design the offer, deliver the offer, deal with customers and support the organization. Business models or strategic innovations provide high risk/high potential opportunities to rewrite the rules and create new competitive space through a combination of radical product/market, value chain and first-mover innovation management.

The resources, the environment and the expectations a firm faces will condition which of these strategic levers it should pull from an innovation point of view, as well as how innovative its strategy should be. Being the most innovative does not necessarily mean being the most successful; each firm must identify and share throughout the organization the specific innovation posture chosen, in particular which types of innovation opportunities it wants to pursue and how much it wants to invest in them.

Further reading

Adner, R. (2006), "Match Your Innovation Strategy to Your Innovation Ecosystem", *Harvard Business Review*, pp. 1–9.

Besanko, D., Dranove, D., Shanley, M. and Schaefer, S. (1996), *Economics of Strategy*, Hoboken, NJ, John Wiley.

Chan, K. (1997), "Value Innovation: The Strategic Logic of High Growth", *Harvard Business Review,* January/February, pp. 103–112.

Chan, K. and Mauborgne, R. (2005), *Blue Ocean Strategy: How to Create Uncontested Market Space and Make the Competition Irrelevant*, Boston, MA, Harvard Business School Press.

Cheng, Y. T. and Van de Ven, A. H. (1996), "Learning the Innovation Journey: Order Out of Chaos?", *Organization Science,* Vol. 7, pp. 593–614.

Chesbrough, H. and Rosenbloom, R. S. (2002), "The Role of the Business Model in Capturing Value from Innovation: Evidence from Xerox Corporation's Technology Spin-Off Companies", *Industrial and Corporate Change*, Vol. 11, pp. 529–555.

Christensen, C. and Raynor, M. E. (2003), *The Innovator's Solution*, Boston, MA, Harvard Business School Press, Ch. 1.

Davila, T., Epstein, M. J. and Shelton, R. (2006), *Making Innovation Work*, Philadelphia, PA, Wharton School Publishing, pp. 75–79.

De Wit, B. and Meyer, R. (2003), *Strategy: Process, Content, Context*, Florence, KY, Cengage Learning Business Press, Ch. 8.

Gluck, F. W., Kaufman, S. P., Walleck, A. S., McLeod, K. and Stuckey, J. (2000), "Thinking Strategically", *McKinsey Quarterly*, June.

Freeman, C. (1988), *The Economics of Industrial Innovation*, Cambridge, MA, MIT Press.

Govindarajan, V. and Trimble, C. (2007), *10 Rules for Strategic Innovators: From Idea to Execution*, Boston, MA, Harvard Business School Press.

Hamel, G. (1996), "Strategy as Revolution", *Harvard Business Review*, July/August, pp. 69–82.

Hamel, G. (1998), "Strategy Innovation and the Quest for Value", *Sloan Management Review*, Vol. 39, pp. 7–14.

Hamel, G. and Prahalad, C. K. (1994), "Competing for the Future", *Harvard Business Review*, July/August, pp. 122–128.

Lieberman, M. B. and Montgomery, D. B. (1988), "First-Mover Advantages", *Strategic Management Journal*, Vol. 9, pp. 41–58.

Markides, C. (1997), "Strategic Innovation", *Sloan Management Review*, Vol. 38, pp. 9–23.

Markides, C. (1999), *All the Right Moves: A Guide to Crafting Breakthrough Strategy*, Boston, MA, Harvard Business School Press.

Moingeon, B. and Lehman, O. (2006), "Strategic Innovation: How to Grow in Mature Markets", *European Business Forum*, Vol. 24, pp. 50–54.

OECD (2009), *Innovation in Firms: A Microeconomic Perspective*, Paris, OECD.

Porter, M. E. (1985), *Competitive Advantage*, New York, Free Press.

Schilling, M. A. (2006), *Strategic Management of Technological Innovation*, 2nd edn, Boston, MA, McGraw-Hill, pp. 83–99.

Skarzynski, P. and Gibson, R. (2008), *Innovation to the Core*, Boston, MA, Harvard Business Press.

Teece, D. and Pisano, G. (1994), "The Dynamic Capabilities of Firms: An Introduction", *Industrial and Corporate Change*, Vol. 3, pp. 537–556.

Zook, C. (2007), "Finding Your Next Core Business", *Harvard Business Review*, April, pp. 66–75.

CHAPTER 4

Nurturing entrepreneurial resources

Short case: boosting entrepreneurship at Plastics Ltd.[1]

Plastics Ltd. is a European chemical company created a century ago. It employs thousands of people in factories and plants across the world and has built its success on revolutionary research-based product and process developments. It still makes a sizeable share of its profit from products it invented decades ago and most of its top managers began their careers with the firm. Hence it has developed over the years an engineer-driven "invented here" culture, based on efficiency, safety, corporate loyalty and commitments to plans and targets.

Faced with global competition, environmental regulations and the increasing commoditization of its product categories, Plastics Ltd. had tried in the past to launch innovative projects. However, these initiatives met with skepticism from managers and often strong resistance across the organization. Indeed, such initiatives involved new, complex processes and often were at odds with the dominant culture and historic corporate values.

This has led the firm to launch major initiatives recently, aimed at accelerating the pace of innovation and in particular attracting, developing and retaining more entrepreneurial resources. These initiatives include both hard and

$\rightarrow$

soft measures, at individual, team, organization and eco-system levels.

To boost awareness and initiatives at the individual level, a corporate university was launched, with dedicated programs covering entrepreneurship, risk management and leadership issues. Individual assessment and reward processes were updated in order to explicitly include criteria directly linked to the identification and implementation of innovation opportunities. Finally, the yearly corporate employee survey now includes several questions linked directly to the attitude and perceptions of employees regarding innovation.

To stimulate innovation and entrepreneurship at team level, a global corporate competition for the best innovation projects is now organized every few years, celebrating the best successes and identifying and valorizing innovative role models. A dedicated new business development structure has also been created, to offer a fast-track process for teams developing promising radical innovations that could not be developed within existing business units.

At the corporate level, a global internal communication plan was launched and the strategic objectives of the business units now explicitly integrate the management of innovations. Moreover, a corporate innovation manager has been appointed (hired from outside), reporting directly to the executive board and supported by a global network of innovation champions, present in all the major businesses and regions. These help to stimulate the diffusion of innovative modes of management and the sharing of experience and best practices across the organization.

Finally, at the ecosystem level, the firm has defined explicit objectives regarding the proportion of innovation projects that should involve partnerships. It has also developed dedicated processes regarding the identification, assessment and

$\rightarrow$

implementation of partnership opportunities, in particular the opportunity to invest in promising new ventures. Finally, the firm also launched a strategic partnership with a business school to develop its corporate entrepreneurship capabilities.

As a consequence, Plastics Ltd. has managed both to improve its internal and external image as an entrepreneurial and innovative firm, and to launch several promising new businesses.

There is no point defining a strategy that your organization cannot implement. Having outlined how innovation should be integrated with the strategy of a firm, we shall discuss in this chapter how it can be integrated into its organization.

We highlighted in Part I that innovation management is about not only identifying new opportunities but also making change happen. And change can happen only if the organization has the right motivations (will) and the right abilities (skill). Such motivations and abilities should be understood and managed at four different levels: the people; the teams; the organization itself; and finally the ecosystem in which it operates. We shall review these four levels below.

Innovative people: nurturing corporate entrepreneurs

Innovation is about change, and change will happen only if some people are ready to take risks and make an impact on their environment. These change agents or corporate entrepreneurs are the people ready to confront the economic and social forces that maintain the status quo in order to create value and self-realize.

Who is an entrepreneur?

Entrepreneurs can be defined as the people who are able to identify new combinations that involve "leaps forward in the face of uncertainty".[2] Entrepreneurs are also those who are ready to "pursue opportunities beyond the resources they currently control".[3] An entrepreneur is "somebody who upsets and disorganizes" (Peter Drucker), "the bold and imaginative deviator from established business patterns and practices" (William Baumol).[4]

Entrepreneurs are often associated with the independent creation of new organizations, involving personal and/or investor's risks. According to popular myths, entrepreneurs are lonely heroes, typically college drop-outs, who raise millions from venture capital firms to develop world-changing products.

But these are, indeed, myths. Entrepreneurship is an intense social activity, involving business partners and personal networks rather than lonely geniuses. Entrepreneurship also requires experience, which most college drop-outs do not have (the average age of a new business creator in the United States is 39).[5] Moreover, most new ventures are financed through personal debts and informal sources of capital, not professional (venture capital) funds. Finally, most new ventures develop new processes and new business models, often adapted from the founder's own experience, rather than revolutionary products.

Entrepreneurship is a process that can be found and nurtured inside existing organizations, by people focused on developing innovations within a firm and ready to face both corporate and personal risks. In this case, the process is not only about organizing and mobilizing resources, but also about pushing initiatives, as activities are often restricted by the organization's routines and procedures.

Such corporate entrepreneurship can be defined as the process whereby "firms extend their activities in areas unrelated, or marginally related to their current domain of competence based on internal developments and new resource combinations".[6] Corporate entrepreneurship can also be defined as "employee initiatives from below in the organization to undertake something new".[7] It typically involves both "formal and informal activities aimed at creating new business within established firms through product and process innovations as well as market development".[8] Finally corporate entrepreneurship can also reenergize and enhance the ability of a firm to acquire innovative skills and capabilities.

But where can we find these corporate entrepreneurs? And how to recognize them? For decades, psychologists, recruiters and human resource managers have attempted to identify the defining traits of entrepreneurs, with very limited actionable results. Entrepreneurs (along with prisoners!) have actually been found to suffer on average more from dyslexia than does the average member of the population.[9]

Entrepreneurs do tend on average to express a higher need for achievement than is seen in the average population. They are unhappy with the status quo. Entrepreneurs also tend to have a rather internal locus of control. They are more confident and they believe that what happens to them depends on their own actions rather than fate. Finally, entrepreneurs tend to express a higher risk-taking propensity than the average in the population. In other words, entrepreneurs, who are defined as people who want to make change happen and confront risks, have been found to embrace change and risk more than the average member of the population.

The conclusion is, to quote a famous entrepreneurship scholar: " 'Who is an entrepreneur?' is the wrong question".[10] People are not born entrepreneurs. Some people will behave like entrepreneurs in some circumstances, as a result of opportunity or (quite often) by necessity.

There is, however, one circumstance that is known to increase significantly the probability that a person will become an entrepreneur: the presence of a role model. If somebody who is close to that person (such as a neighbor, a relative or a colleague) is or has been a successful entrepreneur, there is a high probability that that person will follow in the entrepreneur's tracks.

This means that organizations (or regions) where such success stories are rare or not publicized will tend to be less entrepreneurial. Conversely, it means that generating and publicizing entrepreneurial behaviors in an organization can generate a virtuous circle, as the first entrepreneurs become role models for the following generation. This also means that testimonies from former peers (such as alumni or ex-colleagues) can be much more inspiring than a one-man show of international stars, who sometimes entertain more than they inspire.

Before exploring further what makes or does not make people behave like entrepreneurs, we shall first discuss exactly what this behavior entails, in particular within an existing organization. Who is the real person behind the myths? What does he or she do that others do not?

What do (corporate) entrepreneurs do?

Entrepreneurship today has such positive connotations that, when asked whether they want people in their organization to be more entrepreneurial, most managers will feel pressed to give a positive reply. But when asked what that implies concretely in terms of new behaviors or how they recognize an entrepreneur when they see one, many will struggle to offer more than generalities, such being more dynamic or taking the initiative.

However, corporate entrepreneurs can be identified both in terms of the specific activities in which they engage, and in terms of their distinctive attitudes. We shall describe both below.

On the one hand, corporate entrepreneurs tend to engage in other activities as well as those related directly to the tasks or objectives they have been assigned. Corporate entrepreneurs tend to engage, formally or informally, in four simultaneous sets of activities:

- Identifying new business opportunities (new products, new processes and so on) beyond the current scope of the organization.
- Assessing whether such business opportunities are worth pursuing, combining personal experience or instinct with the facts and figures they can gather.
- Launching the exploitation of those business opportunities, sometimes bending corporate rules and crossing formal organizational boundaries to do so.
- Enrolling people and mobilizing resources, convincing colleagues and upper management to support, or at least to tolerate, their activities.

On the other hand, corporate entrepreneurs tend to adopt distinctive attitudes regarding how they are supposed to do their job. First, corporate entrepreneurs are certainly not completely asocial but they will often behave more as stand-alone, autonomous individuals than as elements of an organization. For example, most corporate entrepreneurs do not see themselves defined mainly by status, a title or a job description.

Second, while more administrator-type managers mainly see their mission as allocating the resources under their control effectively (through, for example, planning, budgeting and controlling), corporate entrepreneurs will tend to focus first on potential opportunities, dealing with the resources needed only *a posteriori*. In other words, they see the need to mobilize resources beyond those that they initially control, if any, as a business-as-usual activity rather than as stepping on to someone else's turf.

Finally, more administrator-type managers will see best practices as acting systematically, according to existing systems, procedures and policies, following rules and benchmarking themselves against others. Their objective is, implicitly or explicitly, to sustain their organization. In contrast, corporate entrepreneurs will be ready to bend rules, act fast, proactively and in a persistent way to create new coalitions.

Obviously, the activities and attitudes listed above refer to archetypes whose individual features can be found to some degree in most people. However, identifying them allows us to define explicitly what types of behaviors and attitudes should be encouraged when managers seek to foster corporate entrepreneurship.

One particular challenge for corporate entrepreneurs is that they must often alter the institutional context in which they operate. As they identify, assess and exploit opportunities, they must deal simultaneously with formal and informal networks of actors and their representatives (customers, shareholders, products or technologies). As a consequence, corporate entrepreneurs have to develop political tactics to identify the resource bottlenecks, and build enrolment and motivation through a range of social exchanges. They have, consciously or unconsciously, to understand which parties cooperate or compete with their projects, as well as who can regulate, conflict with or accommodate the development of these projects.

Hence coalition building and networking, though not often recognized (or encouraged) by managers, are key activities for corporate entrepreneurs, who not only think but also act outside the (corporate) box.

As an illustration, in the famous Post-it story at 3M, the corporate entrepreneur initially faced heavy skepticism from his management. The innovation emerged only because the corporate entrepreneur found ways around that initial opposition, among others by building a coalition of enthusiastic users among his close colleagues. This coalition then helped to convinced management that the initial invention had the potential to become a successful innovation.

Having reviewed the behaviors and attitudes typical of corporate entrepreneurs, we shall discuss in the next section ways to stimulate and encourage those behaviors and attitudes in organizations.

Should they stay or should they go – how to foster entrepreneurs

The managers who believe some people are born entrepreneurs (see above) consequently complain that they do not have such people in their organization, and envy the successful start-ups they sometimes see emerging around them. However, they forget that many of the founders of these start-ups used to work in organizations like theirs. Indeed, entrepreneurship is "a process of becoming rather than a state of being".[11]

It is therefore important to understand which are the drivers of entrepreneurial intention; that is, what can trigger people to want to act entrepreneurially. Let us stress that in many cases the triggers might be related to negative necessity or threats (to survive or to escape poverty), not only to positive motivations (such as seeking wealth or self-realization).

The antecedents of an intention such as wanting to act like an entrepreneur can be grouped into three categories:[12] the attitude toward the entrepreneurial behavior; the subjective norms regarding that behavior; and the perceived behavioral control. We shall discuss these three factors and their implications below.

First, the attitude toward the behavior refers to the degree to which a person has a favorable or unfavorable evaluation or appraisal of the behavior in question. In other words, does the person personally believe that this is the right thing to do? People will behave like entrepreneurs if they perceive that the expected benefits outweigh the expected costs (such as the risk of losing one's job, reputation or wealth).

Second, the subjective norms refer to the perceived social normative pressures to perform or not the behavior – the subject's perception of other people's opinions of the proposed behavior. In other words, does the person believe that other people who are important to him/her, would approve or support the behavior? Entrepreneurs are more likely to emerge in a social environment that recognizes and values their contribution.

Finally, the perceived behavioral control refers to the perceived ease or difficulty in pursuing the behavior. Individuals usually elect to adopt behaviors they think they will be able to control and master. In other words, does the person perceive that she/he would be able to adopt the behavior if s/he wanted to? People are unlikely to act as entrepreneurs if they feel powerless and/or if entrepreneurship is perceived as an unattainable goal. This last factor can be influenced particularly by the presence of role models, who can make becoming an entrepreneur appear more achievable.

Hence there are two aspects related to will (the person's own will and the will of others who are important to him or her) and one related to skills. These aspects have been shown to influence significantly whether a person will adopt behavior such as acting like an entrepreneur.

Understanding these factors is the first step, but what matters for managers is that these three antecedents can be measured (for example, through corporate surveys) and acted upon.

The attitude toward the entrepreneurial behavior can, for example, be influenced by prevailing tendency to risk aversion, by existing incentive systems, and by the relative awareness and perceived sense of urgency among staff members. In particular, managers often communicate officially that risk-taking and initiatives are encouraged. However, their staff might perceive risk-taking as the opposite of safety (for example, in industrial settings, where safety failure might mean casualties). They might resist what can be perceived as dangerous gambles that could expose them to negative impacts.

Subjective norms can be influenced by "walking the talk" – regarding, for example, tolerance of failure or mistakes, and support for risk-taking and change. It can be influenced by the presence of role models (see above) and by the celebration of their successes. Finally, it can be influenced by perceived recruitment, promotion, evaluation and job assignment practices.

One commonly used method to boost these attitudes and subjective norms is to implement employee-level innovation metrics (to frame norms) and rewards (to influence attitudes). These metrics may be based, for example, on the number of publications or patents, the number of ideas generated, performance improvements, competencies gained, or other more qualitative innovation ratings. Rewards can include financial remuneration, promotion or other perks, or may be based on softer recognition benefits, such as praise and publicity, symbols or a plaque, peer recognition or a boost to self-esteem.

Let us stress that, while they provide a "what-gets-measured-gets-done" comfort to some managers, the effectiveness of such employee-level innovation incentives is very context-specific, depending on the functional aspect of the organization and its culture. Moreover, they often generate perverse effects, such as individualistic behaviors and agency costs (agents pursuing their own personal interests rather than those of the organization).

Finally, perceived behavioral control can be influenced by the availability of skills and training, and the access to corporate knowledge and expertise; by team dynamics; and by the level of stress and collaboration prevailing in the company (see next section). In particular, executive training and professional development programs can address explicitly the skills needed to engage in the identification, assessment, launch and enrollment of business opportunities, such as business planning and coalition-building skills.

In other words, whether people will want to act like entrepreneurs in an organization can be analyzed, and various managerial levers can be used to improve it in a systematic and proactive way.

Having discussed the development of entrepreneurial resources at the individual level, we shall discuss in the next section the characteristics of innovative team dynamics.

Innovative teams: building winning packs

The diversity and depth of skills involved in innovation projects, and the level of energy required to implement them means that innovations are generally the products of teams, not single individuals. Those teams can either be informal, loosely coordinated by the entrepreneur, or formally structured, with a dedicated project team, fixed deadlines and resources explicitly allocated.

However, as all sports fans know, a team of stars is not necessarily a star team. It is therefore important to identify the characteristics of innovative teams, beyond the fact that they should include entrepreneurial people (see above). This means understanding what will make those teams effective, innovative and proactive.

Effective innovation project teams

Innovation projects are by their nature rather complex. They involve multifunctional aspects (such as technology, regulation, market or finance) and significant interdependencies between various tasks and people (for example, R&D and marketing). The tasks themselves are often complex, ambiguous and uncertain, as often only limited benchmarks or experience is available. Finally, innovation projects must deal with constrained goals and deliverables, as time and money are in most cases important issues.

As a consequence, innovation project teams cannot escape the discipline of effective project management, which includes:

- An effective project leader (see below) and sponsors (godfathers) as well as core and associated members with clear roles, aligned with an individual behavioral style. In particular, team members should be selected on the basis of their expertise and skills rather than the fact that they happen to be available at the time of the project launch.
- Clearly defined lifespan (project kick-off, milestones and closing) and deliverables, with a constant focus on learning and insights (see the stage gate discussion in Chapter 7).
- Dedicated resources and coordination processes, involving effective conflict-resolution mechanisms.
- Ability to interact with external partners, either in other parts of the organization or third parties.

In particular, each organization must consider how much it wants to standardize the project management tools and processes used across its innovation projects. More standardized approaches will allow the organization to deal with more numerous, bigger and/or more complex projects, and build synergies between them. Adopting standardized approaches can also provide legitimacy for the projects *vis-à-vis* the rest of the organization.

But very effective teams using standardized approaches can sometimes become armies of robots or flocks of sheep. They ultimately become "exactly wrong rather than approximately right". What will make such teams not only effective but also innovative is discussed below.

Innovative project teams

What makes some teams highly innovative even when, at least on paper, their members are not expected to be innovative, has been the subject of intense research. While the level of pressure the team has to deal with certainly matters (see next section), how the team functions – its dynamics – also plays an important role.

Whether teams will work like "sparkling fountains or stagnant ponds"[13] will in particular be driven by, on the one hand, the match between tasks and people, and on the other, the interactions between team members.

The right task for the right person means first that the team job descriptions should generate the right challenges and foster group task orientation. Innovative teams need great autonomy of means and clear accountability, in particular regarding their tasks and objectives. The tasks involved should also engage and interest the team members and have strong collaborative implications, without being too difficult or diversified. In other words, the aim is to develop commitment, not merely assignments.

Second, the team composition should provide both the right knowledge or expertise and the right level of diversity. Innovative team members should be selected based on their personality (are they sufficiently likely to behave like entrepreneurs?) and the background they bring to the project. This includes not only their technical, procedural or intellectual knowledge, but also their life experience and social networks. Innovative teams should have access to different sources of information and expertise.

But team members should not all have a similar background and experience, functional skills or even age. Teams should involve newcomers whenever possible. Innovative teams combine different pools of contacts and different creative thinking skills, such as left brain versus right brain, diverging versus converging, or storming versus norming.

One should, however, not stretch diversity too far, as team members should still respect and understand each other. This also means that the group diversity must be dealt with, by providing safety and the right level of integration and bonding to all members, to avoid the risk of ostracism. Indeed, insufficient diversity hinders positive friction and the cross-fertilization of ideas. However, too much diversity creates chaos, fragmentation and frustration.

On the other hand, managing the interaction between team members first means creating an energizing and demanding group dynamics. This is achieved mainly through the right level of supervisory management. In particular, the level of pressure and uncertainty must be stimulating without being paralyzing (see below).

Second, the way the team balances alternatives and makes decisions is also critical. An effective team leadership (see below) can build consensus while escaping the trap of groupthink, by combining both diverging and converging team thought processes and productively managing conflicts. The team must also avoid excessive social identification ("us versus them") and maintain its ability to liaise with the outside world and integrate external input. They should be ready to accept "proudly found elsewhere" projects rather than to reject them because they were "not invented here".

This is particularly the case when virtual teams are involved (through, for example, teleconferencing or online collaborative tools). In this case formal decision processes and communications often prevent differences or growing discomfort to surface. Nothing yet beats periodic face-to-face interactions when subtle decisions have to be made and when enrolment is key, in particular when cross-cultural differences are involved.

One consequence of the need for effective and innovative project teams is the challenge for innovative organizations to maintain and develop a pool of talented innovation project managers. As team architects, these managers must be open-minded and optimistic enough to be able to act as leaders, role models and group facilitators. As network builders, they need to be able to provide advice, feedback, criticism, rewards and support while always maintaining their integrity. They must also create institutional spaces where team members feel empowered, trusted and free to experiment. Finally, as resource jugglers, these managers need strong interpersonal skills to be accessible, to listen and to create coalitions. Overall, they need to be able to provide extrinsic and intrinsic motivations to make things change. Indeed, as M. P. Follet found out (in 1924), "Leadership is not defined by the exercise of power but by the capacity to increase the sense of power among those who are led. The more essential work of the leader is to create more leaders."[14]

One key challenge for innovation project leaders is to find out how much pressure to put on team members. This will be discussed below.

Proactive innovation teams: stretch versus stress

A common complaint among employees or managers is that they are fully ready to innovate, if only they had the time. For them, the uncertainty and delayed benefits associated with innovation combined with the pressure of day-to-day business often hinder investment in innovation.

As one way of dealing with this problem, some organizations (such as 3M or Google) create spaces and periods of time where employees are free to experiment and where innovations can flourish. However, there is very limited evidence available regarding the direct positive results of such an approach.

So what is the right level of stress? The answer is that, while very low or very high stress is good for creativity, it is often bad for innovation.

Low stress might allow the mind to wander freely, boosting creativity as in the proverbial idea hit upon while taking a shower or day-dreaming. But low levels of stress also foster lethargy, apathy and a low motivation to make change happen. For example, many R&D labs in the 1970s were perceived by corporate managers to be unmanageable black boxes. They were indeed often very creative. But from Xerox to Kodak, IBM and Philips, many failed at that time to manage some important innovations effectively.

Indeed, if there is limited perceived urgency to change, there will be little incentive to bear the costs and risks of innovation. This is one of the reasons why some corporate dinosaurs who believed they would be around for ever failed to react to innovations that ultimately destroyed them (see Chapter 2).

On the other hand, high stress can create burning platforms that might stimulate the urge to find new ideas and new ways to operate (as in the movie, *Apollo 13*). However, too high a level of stress is known to generate conformist behaviors, as people curl up and retreat into familiar modes. High organizational stress can create an inability to explore and assess opportunities and to mobilize resources to capture them, as urgency blinds decisions and no bandwidth is left for new opportunities to emerge.

For example, during financial crises some bank managers, regulators and policy makers have been paralyzed by the scale of the complexity, risks and dangers all around them. Hence stress might create the motivation to change but too much stress can destroy the ability to make that change happen. Innovative team leaders are those who achieve the right balance between those two extremes.

Having discussed what characterizes innovative people and teams, we shall discuss in the next section the organizational features that can stimulate them or provide a hindrance.

Innovative organizations: deliverables, cultures and structures

In the same way that here are some gardens where across the seasons trees are bigger and flowers are more numerous than in other gardens, there are also organizations where across business cycles innovative teams and people are more prevalent and more successful. It is therefore important to understand what characterizes these innovative organizations. What are the things they do better than others? How is that influenced by soft or cultural factors and by hard or structural ones? We address these issues below.

Innovation metrics and deliverables

The first thing to agree upon is what is meant by "innovative organizations". In other words, what are the yardsticks or metrics to track and/or the deliverables to expect from these organizations, taking into account their strategic objectives and the performance of their competitors.

Indeed, there appears to be no single accepted way to measure the innovativeness of an organization. Many metrics and indicators have been proposed and used to measure and benchmark the innovation performance of an organization, in terms of inputs, processes or outputs.

Input metrics include indicators such as R&D and innovation spending, the number of innovations ideas or projects initiated, the proportion of personnel involved in innovation projects, and the number of partnerships or the number of performance gaps identified and acted on.

Process metrics include project costs or outsourcing efficiency, time to market or time to break even, or the proportion of projects stopped or on hold. It can also include the number of patents or the rate of adoption of advanced innovation techniques or tools. Finally, it can also refer to the breadth and depth of the open innovation network of the organization (number and quality of partners).

Input and process metrics are popular with economists and consultants because of the benchmarking and statistical opportunities they provide. However, they can be very dangerous for managers to use. First, they tend to encourage innovation as an objective *per se* (what gets measured gets done), rather than as a means of achieving the objectives of the organization (see Chapter 2). Second, it is very difficult to find indicators that are directly comparable from one firm to another, in particular when comparing firms of different sizes or from different sectors.

For example, an expense that is included in the R&D budget of one firm might be considered as technical support, maintenance, customer service or an engineering expense in another (in particular, for firms where R&D expenditures are publicized). Similarly, the number of patents generated by a firm can be as much influenced by its innovation performance as by the patenting and secrecy tactics it uses or the regulatory environment it faces. For example, the USA, Japan and European Union patent offices tend to deal with the same patents in different ways.

Finally, the R&D expenditure of a corporation or a region (a very popular benchmark among policy-makers and economists) is influenced strongly by its industrial structure – for example, the relative importance of service activities, regardless of how well innovation is managed there.

Input and process metrics should therefore be used with caution. They can, however, provide useful soft targets aimed at raising the awareness and involvement of people across the organization. Examples of such soft targets include expecting one idea from each employee or demanding that 50 percent of innovation projects involve partners. But blindly tracking these metrics can lead to counter-productive decisions, such as generating or patenting ideas just for the sake of it.

As a consequence, it is more effective, whenever possible, to use output targets aligned with the strategic objectives of the organization. Those include sales from new or improved products, market share growth or operational improvements. They also include the return on investment (ROI) of innovation projects or the development of intellectual property (IP) revenues and of an innovative image. Finally, they can relate to skill building (such as the number of people trained) or strategic value (such as the size and quality of the pipeline of innovation projects).

While such *ex-post* input, process or output indicators can allow managers to track and in some case benchmark the innovation performance of an organization as a whole, they tend to be difficult to use as operational yardsticks for people and teams across the organization. For example, a sales person will in general not be motivated by the number of patents. An R&D expert will sometimes feel very little connection between his or her personal contribution and the evolution of the market share of the business.

Hence another approach is to define *ex-ante* deliverables that the organization wants to achieve across its teams and people. They can be categorized either in terms of continuous improvement (doing the same things better), new innovation mindsets or attitudes (doing things in a different way) or new business developments (doing new things). We discuss the latter two categories below.

In particular, the innovativeness of an organization can be improved by raising awareness and developing new mindsets regarding innovation. In that case, teams and people across the organization are encouraged to act in a more proactive and persistent way, in order to take ownership and learn new things. They are also encouraged to focus on pursuing opportunities by taking new initiatives and being more open-minded and/or market oriented. People and teams can also be encouraged to deal more proactively with risk, raise their personal commitment and become more autonomous by tolerating ambiguity and uncertainty, challenging routines and mobilizing resources.

This *new mindset* approach is relevant in particular when pursuing exploitation strategies (see Chapter 2) and incremental innovations.

From another point of view, the innovativeness of an organization can also be improved by raising its ability to identify and capture new opportunities, outside its current scope of activity. This means building the capability to pursue somewhat high-risk but also high-potential projects (play poker, not chess), often through dedicated teams and people. It implies it will be ready to develop new product ranges, enter new markets or build new value chains, as long as the firm can leverage its scale, unique assets or reactivity. It also means being able to manage resources and time horizons beyond the scope of current business units and processes (such as annual budgets and quarterly earnings), while stopping early costly irrelevant ventures or pet projects.

This *new business* approach is particularly relevant when pursuing exploration strategies (see Chapter 2) and radical innovations.

Let us stress that both exploitation (new mindset) and exploration (new business) perspectives are aimed at making the organization more innovative, but that they differ radically in terms of implementation and expected outputs. Again, rather than wanting to be innovative per se, what matters is to clearly define what a firm wants to do and why.

Having discussed how the innovativeness of an organization can be measured and defined both *ex post* and *ex ante*, we shall discuss in the next section some of the key soft or cultural factors that matters, in particular when considering the new mindset perspective.

Innovative climate: fostering new mindsets

In the first section of this chapter we highlighted the importance of attitudes and social norms as well as the statistical significance of the presence of role models regarding whether or not individuals adopt entrepreneurial behaviors. For organizations, this means that shared beliefs or assumptions about how their members perceive, think and feel in relation to entrepreneurial opportunities – in other words its innovation culture matters strongly.

Therefore, how members of an organization perceive "how things are done here" regarding knowledge-sharing and donating, bending rules or dealing with uncertainty will have a strongly influence on whether and how effectively they will embrace innovation opportunities and adopt new mindsets. This concerns, in particular, how people trust each other and share a collective responsibility rather than compete against each other. It also concerns how much people can take risks and are able to learn from mistakes. Finally, this concerns the accessibility of the management and how it interacts with the rest of the organization.

Indeed, unsupportive culture and climate are often quoted by CEOs as the Number One internal organizational barrier to effective innovation,[15] in particular regarding the ability of projects to fail without penalties and the lack of coherent strategic vision.

Whether organizations, and in particular their management, can adopt a strong innovation culture will depend on the external pressure they face (see Chapter 2), but also upon several formal and informal internal factors:

- The leadership style and the strategy, leading to clear priorities, capabilities and involvement from the management as well as a readiness to take risks and nurture innovation projects. In particular, the willingness of senior management to facilitate and promote entrepreneurial activity in the organization is

key, including championing innovative ideas as well as providing necessary resources, expertise or protection.

- The formal policies and standards adopted by the organization
  and the systems that support them – for example, regarding the
  allocation of resources, responsibilities and rewards as well as
  the management of partnerships and IP or the management of
  knowledge and ideas.
- The formal rules in place regarding decision-making processes and responsibilities. This includes how the organization
  is structured (such as layers, boundaries and reporting channels), how much it is centralized, and the level of freedom and
  autonomy available. This also relates to whether formal stage
  gate or equivalent processes such as project management platforms are in place (see Chapter 7).
- The profile of the people recruited (staff and skills), their education, background and training, and the existing communication channels.
- The informal characteristics prevailing, in terms of shared
  values, climate and dominant social networks.

The challenge for managers is therefore to develop a coherent
approach regarding these levers in order to foster an innovation
culture. This can be done by the managers directly and/or by
appointing dedicated support resources such as a network of
corporate innovation champions.

Conversely, the development of an innovation culture may be hindered by a combination of structural and behavioral organizational
barriers, such as:

- A predominance of restrictive vertical relationships (organizational silos not communicating) and a bias toward incremental innovations ("competency trap" or lack of readiness
  to explore new areas).
- Risk-averse planning and unsupportive accounting practice,
  such as the exclusive use of return on capital metrics or a focus
  on short-term efficiency rather than portfolio value.
- A limited availability of effective tools and resources and
  half-hearted funding constrained by annual budget cycles or
  bureaucratic stage gate processes.

- Negative response from management (sense of complacency, criticism, feeling of threat), driven by "not-invented-here, we-have-always-done-it-this-way" syndromes or a culture of inferiority.

How well managers combine these factors and deal with the barriers in order to create an innovation culture can be assessed, among other ways, by evaluating the "entrepreneurial orientation" of the firm.

The entrepreneurial orientation of a firm measures how effective are the processes, practices and decision-making activities of its management, regarding five dimensions of innovation management that have been shown to be linked with business performance and growth:[16]

- Innovativeness: a willingness to support creativity and experimentation in introducing new products or services as well as novelty, technological leadership and R&D in developing new processes.
- Risk-taking: a tendency to take bold actions such as venturing into unknown new markets, committing a large portion of resources to ventures with uncertain outcomes, and/or borrowing heavily.
- Proactiveness: an opportunity-seeking, forward-looking perspective involving the introduction of new products or services ahead of the competition, and acting in anticipation of future demand to create change and shape the environment.
- Autonomy: independent actions led by individuals or teams aimed at developing a business concept or vision and carrying it through to completion.
- Competitive aggressiveness: the intensity of a firm's efforts to outperform industry rivals, characterized by a combative posture and a forceful response to competitor's actions.

Hence each management team can review how it behaves and performs regarding these five dimensions. It should also consider whether the resulting "innovation posture" is aligned with the company's strategic objectives. This is particularly true for large organizations, where there can be large discrepancies between top-management vision and front-line action. Whether and how large firms can also be innovative is discussed in the following section.

Innovative climate in large firms: can elephants dance?

One common assumption among managers and some policy-makers is that "entrepreneurial orientation" is limited to small firms, supposed by their nature to be more dynamic, while large firms are stuck with inertia and bureaucracy and are therefore less innovative.

There are indeed barriers to innovation that are linked to size. As the organization grows, its members typically experience a loss of managerial control and an increase in bureaucratic iner-tia. As a consequence, potential entrepreneurs might experience a lower individual incentive to innovate, as their perceived abil-ity to make a visible impact diminishes.

Similarly, fears of a threat to, or a dilution of, existing earnings decrease the collective incentive to innovate. Large firms have also over time often built up strategic commitments to employ-ees, suppliers or customers that can conflict with or be canni-balized by innovation projects.

However, one should not forget that large firms also enjoy sig-nificant benefits that others miss regarding the management of innovation. They often have the ability to mobilize resources not available on the market, such as their brand and credibility, their customer base and market power, or slack logistic or man-ufacturing capacities. Large firms also have the ability to spread the risk and fixed costs of innovation (in particular R&D) over larger sales volumes, and to capture scale and learning effects across their portfolio of projects.

Finally, large firms typically also enjoy a more global reach and more favorable access to complementary capabilities, in terms of technology, routines, cash or people. As an example, any mom-and-pop shops can try to discover a better recipe for cola, but that does not mean that large numbers of people will stop drinking Coke or Pepsi.

As a consequence, both small and large firms can be in a position to manage innovation effectively. Smaller firms will, however, in general focus on niche, less mature and more risky projects, where experimentation and flexible decision-making matter, because they have less to lose. In other words, they are often better armed to deal with the *newness* aspect of innovation.

Conversely, large firms will often be more effective at later stages, when manufacturing and marketing effectiveness as well as scale matter, because they can mobilize resources more effectively. In other words they are often better armed to deal with the *change* aspect of innovation.

This is the case, for example, in the pharmaceutical or software sectors, where the discovery and testing of new products is increasingly the domain of small firms and independent labs, while the big players focus on the manufacture, marketing and distribution of new but validated concepts.

The challenge for managers is therefore to try to balance the existing management style with building an "ambidextrous organization",[17] able to combine the discipline and efficiency of traditional management in large firms with the reactivity and agility of small innovation ventures.

At one extreme, a disciplined or mechanistic management approach focuses on the fit, alignment and continuous improvement of its activities, and on the consistency, reliability and accuracy of the information it collects. It does so in order to analyze, plan and control resources and jobs based on expected returns and volume.

This mechanistic approach may be associated with an industrial mode of management,[18] where hierarchy runs deeply into the organization, leadership is centralized and work is segmented. In such a model, people are essentially considered to be opportunistic individuals and cost centres, which are controlled and incentivized through authority. Such an approach is well suited to exploiting strategies (see Chapter 3).

Conversely, at the other extreme, a flexible or organic management approach can deal with complexity, revolutions and constant shifts in cognitive models and pictures of what the future could be. It aims at breakthroughs involving diversity, dynamism and ambiguity, which require experimentation, testing and uncertainty management. It does so in order to make adjustments to the way its people behave and maximize its success rate.

This organic approach may be associated with an "expert"[19] mode of management, where the hierarchy is quite shallow, leadership is distributed and work is essentially collaborative. In such a model, people are considered to be valuable assets, controlled and incentivized mainly through influence and role models. Such an approach is well suited to exploring strategies (see Chapter 3).

Obviously, these are two theoretical archetypes representing extreme cases. The key for an innovative organization is to capture the opportunities of a more organic approach while keeping the strengths of a more disciplined one. Where the right balance is for each firm will depend on its strategic objectives (see Chapter 3) and on the nature of the organization, from simple or ad hoc structures (start-up or project-based firms) to machine bureaucracies and division-based corporations. Each firm should try to design the combination of mechanistic and organic management or "innovation archetype"[20] that better suits its strategy.

A complementary approach to the development of an innovative climate or new mindset as discussed here, in particular for large firms, is to focus less on the culture of the organization as a whole but rather to create dedicated structures or support systems where innovation can flourish locally. Those will be discussed in the next subsection.

Innovation structure: corporate venturing

Developing an innovation culture across the whole organization through the various managerial levers outlined above represent in many cases a tough and long-term challenge, in particular for large organizations with mature business activities. Furthermore, different parts of the organization (for example, different business units) might face different circumstances in terms of resources, environment and purpose (see Chapter 3) and therefore call for differentiated innovation strategies and organization. As a consequence, fostering an innovation culture across the organization as a whole is not always the best or a unique option.

A complementary avenue for firms is to develop dedicated innovation support structures or systems, where conditions are fine-tuned locally to stimulate the effective management of innovation. In particular, such structures can be designed to foster the entrepreneurial behaviors and team dynamics described above, through dedicated resource and reward management systems.

These structures can be temporary or ad hoc, focused on specific projects through corporate task forces or through the creation of spin-offs (see below). They can also be set up as long-term multiprojects infrastructures – for example, as corporate incubators or new business development entities.

In these structures, dedicated metrics, leadership styles, formal rules and expertise which might be impossible or undesirable to develop across the whole organization can be combined in a way that specifically fits the innovation strategies pursued. For example, large industries or utilities might be ready to experiment and take risks in dedicated circumstances, perhaps focused on developing new businesses. Similarly, firms with a strong risk-averse and/or bureaucratic culture can create dedicated incubator or separate new venture units rather than trying to change the culture of the whole organization. Finally, a product-oriented firm can launch a spin-off entity aimed at developing new value-added services, involving different culture and reward systems.

These structures can be developed within the organization itself (internal ventures) or at arm's length (external venturing). Both aspects will be discussed below.

Internal ventures relate to the creation or development of new businesses within an existing corporate organization. Depending on the innovation strategy pursued, they can be totally embedded in business units or reporting directly to the top management. They can focus on opportunities closely related to the core business and incremental innovations, or try to foster diversification through the development of radical innovations. Finally, internal ventures can emerge either as formal ventures sponsored by the organization (top down) or through informal entrepreneurial efforts and employees' initiative (bottom up).

Internal ventures can be more effective at developing synergies with existing businesses and at leveraging the benefits of the larger size of the parent company, such as economies of scale or learning (see above). Still active within the borders of the organization, internal ventures might, however, suffer from inertia in decision-making and funding, lack of diversity in teams or risk aversion.

To maintain a strong coherence in the management of the ventures and balance those benefits and disadvantages, internal ventures are in general managed by a dedicated high-level steering committee or internal venture board. This committee must, in particular:

• Manage the value of the portfolio of innovation opportunities pursued by the internal venture structure, by monitoring individual ventures and by ensuring the prioritization of resources to the most promising projects in order to avoid dispersion.
• Pick and motivate the managers most likely to adopt effective entrepreneurial and team leadership behaviors as well as ensure adequate staffing.
• Protect the ventures from corporate constraints while maintaining bridges between the ventures and the internal expertise and capabilities of the business units.
• Provide dedicated support to individual ventures regarding issues such as human resources, IP, use of assets or technical expertise.

On the other hand, external ventures result in the creation of semi-autonomous organizational entities that reside outside the existing organizational domain. This can be done through direct or indirect investments in third parties (for example, through corporate venture capital – see Chapter 5) or through the creation of autonomous "spin-off" structures allowing the management of potential innovation outside the company.

External ventures allow an organization to increase its ability to be flexible and respond to local needs from customers, employees, shareholders or partners. It can also be a way to leverage specialized knowledge or market contacts. However, the "outsider" status of external ventures can generate transaction costs with the parent company, hindering their ability to leverage capabilities such as critical operational skills, manufacturing and distribution assets or support with overheads. Finally, creating a separate entity might protect the brand and credibility of the parent company (in particular in high-risk projects) but conversely can also prevent the new venture from taking advantage of these benefits.

In the specific case of a spin-off, the creation of an autonomous entity can allow a better alignment of incentives in order to retain valued employees – by using stock options perhaps – and the implementation of adequate organizational settings – for example, by using different office locations. It can also make it easier to share knowledge, risk and investments with third parties without involving all the assets of the parent company. Finally, it can be a way of divesting non-core activities while still capturing some of the value created – in particular when the new venture enjoys growth opportunities that generate high market valuations.

Internal and external venturing initiatives might seem very attractive on paper and have been experimented with by many organizations, especially in economic growth periods when shareholder expectations are high and risk aversion low. However, the effective implementation of them generates strong challenges that make them very vulnerable, in particular to business cycles, new CEO priorities or changes of mood among employees or shareholders. These challenges include:

- Organizational inertia, as an overall environment hostile to creativity overwhelms concrete efforts to change, and as organizational rigidities and various bureaucratic obstacles lead to a lack of real commitment to new ventures.

- Lack of entrepreneurial skills, as the organization fails to develop and retain employees and managers with the right skills, in particular regarding the management of multiple products in rapidly changing markets.
- Corporate "short sight", as the firm focuses on historical sources of innovations. In that case, the scanning and recognition of new ideas and opportunities is limited to traditional sources (such as R&D and marketing) and innovation activities can become routine operations.

In summary, organizations must create environments where people are likely to adopt entrepreneurial behaviors, and where innovative team dynamics is fostered. This can be done by encouraging new ways of doing things – new mindsets – and a new climate across the organization, even if the organization is large. It can also be done by creating dedicated structures or systems where local conditions are fine-tuned to foster innovation, in particular the ability to set up new things – new businesses – and develop new ventures inside or close to the organization.

But in the same way that people and teams operate in environments that affect their ability to innovate, so do organizations. How the environment or innovation ecosystem where the firm operates affects its ability to innovate will be discussed in the next section.

Innovative ecosystems: networks of innovation

Firms do not compete with each other in a vacuum. In particular, the way they are able to manage innovation will be influenced by the local environment in which they operate – their innovation system, and by their interactions with peers – their innovation networks – within their sector, their value chain or based around specialized themes. We shall discuss both aspects below, and introduce the notion and challenges of open innovation.

Innovation systems

The characteristics of the socio-economic environment in which the firm operates strongly influence its ability and readiness to manage innovation, with regard to both the prevailing institutional arrangements and the local resource endowments.

The institutional arrangements relate first to the local regulatory and policy context. It concerns the management of property rights, in particular regarding IP but also investments and facilities, and how effectively these rights are defined and protected in practice. It also concerns local workplace and product regulations and norms, which might constrain how and where a firm is able to innovate. Finally, it conditions the firm's legitimacy and right to operate in its environment.

Institutional arrangements also relate to prevailing technical standards and market mechanisms, which define how goods can be made, distributed and sold. They can, in particular, strongly affect the rate of adoption of an innovation (see Chapter 1).

On the other hand, the local resource endowment relates to local or available suppliers, customers, investors, employees and technologies. Local suppliers will influence input prices and available resources (for example, for raw materials or energy supply) while local customers' taste and needs will affect output prices. Locally active investors will influence both the risk profile and the funding of innovation projects, while the local human capacity will constrain the pool of competent human resources available. Finally, local technological sophistication will constrain the pool of available scientific and research competencies.

Taken together, these aspects will define the local innovation infrastructure that the firm can leverage or accommodate, and will affect the way it can develop knowledge, compete and create value. It is, for example, possible to rank countries according to their economic performance, government and business efficiency as well as by the quality of their infrastructure; and to adjust the innovation strategy of the firm accordingly.

Adapting the way innovations are managed around these aspects is therefore critical, in particular for firms operating in environments with which they are not familiar, such as the subsidiaries of multinational companies active in emerging countries. While this is true for management processes and systems in general, it is particularly the case when considering innovation management and its social, cultural, technical and regulatory dimensions.

How this innovation infrastructure affects the emergence of innovations and how it can impact on economic development and wealth creation has led many policy-makers to intervene proactively to attempt to manage this infrastructure. There is therefore an ongoing debate about how much the state should intervene regarding innovation, beyond enforcing basic rules of law such as property rights, and providing basic services such as education or communication infrastructures.

But public authorities have sometimes had a bad track record regarding their ability to anticipate innovation, define in advance which technology path should be explored, and in particular to "pick winners". This can be linked in some cases to the complexity of the expertise required, but also to the tension between the intrinsic uncertainty and fast pace of innovation and the relatively slow pace of policy-making. By the time priorities are negotiated, agreed on, translated into legal frameworks and implemented, and all stakeholders involved have adjusted their strategy, new knowledge that questions the initial assumptions often emerges.

There is, however, a strong case that, rather than trying to pick winners, public authorities have an important policy role to play regarding the development of innovation. This strong case relates to the existence of market failures and positive network externalities linked with innovation. It motivates the implementation of public support to innovation at regional and international level.

Market failures arise when dealing with innovation opportunities because innovation projects are not simply goods that can easily be traded. Innovation opportunities often involve the development of knowledge that is a public good and cannot be appropriated easily (hence the need for IP rights and confidentiality or exclusivity agreements). Innovation also typically involves high uncertainties and irreversible commitments to specific assets which generate risk aversion, agency costs and information asymmetries. Finally attractive innovation opportunities can relate to non-solvent markets – for example, in the healthcare or environment sectors. Those market failures imply that, without adequate public intervention, some attractive innovation opportunities would not be pursued by autonomous economic agents (investors or managers) and therefore could be lost.

On top of these market failures, innovation opportunities can also involve positive externalities. The emergence of innovation contributes to welfare, as problems are solved and new needs dealt with. Furthermore, independent innovation projects can benefit from economies of scope and network effects if they are coordinated or active on the same markets. Finally, the way an innovation is managed by one firm can have important positive spillovers for other firms, through, for example, the development of associated products and services or complementary assets. Those positive externalities imply that some innovation opportunities that would have a positive socio-economic impact would not be pursued by autonomous economic agents, which are assumed to take into account only their own direct costs and benefits.

As a consequence, policy-makers have put in place a wide range of innovation support mechanisms, with varying levels of success in terms of take-up, effectiveness and positive impact. Those support mechanisms may be targeted at individuals, prospective entrepreneurs or specific social groups such as students, unemployed people or specific employment fields (engineers, scientists and so on). They may also be targeted at existing or emerging ventures, either high-potential small businesses such as start-ups, "gazelles" (fast-growing firms), spin-offs or more mature firms considering investment in innovation.

The innovation support mechanisms put in place by public authorities include financial support, either direct (such as grants, subsidies or loans) or indirect (such as market-making initiative or regulatory incentives). They also include logistics support, through the provision of infrastructure (such as science parks or research laboratories), expert services (in fields such as

IP management or exports) or network-building initiatives (such as directories or special events and fairs). They can also involve targeted advice, information or training services, either online or through dedicated programs.

While those support services might involve significant search and administrative costs and match more or less the needs of a firm, it is important for innovation managers not only to know about and use them, but also to influence their design, development and management – for example, through direct contact with policy-makers or through professional associations.

Having outlined the importance of the general infrastructure or innovation system in which firms operate, we shall discuss in the next section the specific innovation networks they can leverage.

Innovation networks

Whether and how much innovation a firm can manage will be influenced by actors beyond public authorities and direct business relations (suppliers and customers). Indeed, the emergence and development of an innovation is in most cases the product of a network of actors, small and large, public and private rather than of a single organization. The importance and quality of that innovation network, as well as the ability of the firm to proactively engage it, are therefore critical elements of the firm's innovation ecosystem.

These innovation networks can include large firms, SMEs and start-ups as well as universities and public research centers. They can also rely on specific key individuals (experts and/or influencers) and might involve dedicated sources of finance such as local banks or specialized investors.

Formal or informal innovation partnerships and collaborations are therefore developed increasingly by firms. Such partnerships and collaborations can indeed generate opportunities:

- To share costs and risks, which is particularly important in view of the rising costs of technology development – for example, in the automobile sector.

- To save time, accelerating product development and the time to market. This is especially important because of the shorter product lifecycle being experienced by many industries – in electronics, for example.
- To better monitor market trends and access international markets, such as through privileged relationships with selected distributors; the fashion industry is an example of this.
- To access specific best-in-class expertise, which would be impossible or costly to develop in house – for example, in the pharmaceutical industry.
- To contribute to the joint development of industrial standards, such as in the information technology sector.

Indeed, between 20 percent and 40 percent of firms across OECD countries report formal and informal collaborations on innovation.[21]

The value of these innovation networks is seen particularly in technology clusters. These are local clusters of firms that have a connection to a common technology and may engage in buyer, supplier and complementary relationships, as well as in research collaborations. Frequent and close interaction between the members of such innovation networks facilitate the transfer of complex and/or tacit knowledge and the development of trust. The repeated nature of the interactions and the cultural proximity of the network members also decrease the likelihood of opportunistic behaviors such as free-riding.

But innovation networks can also take other forms, relying on various types of formal and informal ties, either pre-existing (such as alumni societies or expatriate communities) or ad hoc (for example, professional associations). These forms can vary in terms of their "embeddedness"[22] and degree of purpose, from closed and relatively informal clubs to contractual project consortia, and from highly fluid groups sharing interests to formal strategic alliances.

The increasing importance of leveraging these networks has led to the emergence of dedicated innovation management processes and methods, which can be categorized loosely under the concept of open innovation, discussed below.

Open innovation

Most firms now realize that in a global and fast-evolving environment there will always be more competence and ideas outside their walls, whatever their size and expertise.

First, markets are in general more efficient than individual organizations at constantly creating and destroying potential innovations. The innovation ecosystem of a firm therefore offers ways to diversify and enlarge its sources of innovation opportunities.

Second, innovations often require the development of new market and/or technology capabilities not available within the firm. The innovation ecosystem of a firm therefore offers opportunities to access best-in-class operators and outsource part of the management of an innovation.

Third, the whole process of turning an idea or a potential opportunity into a profitable business requires multiple skills and capabilities at different stages and often over a significant period of time. Indeed, it can easily take five to ten years to go from a business concept on paper to a sizable and profitable business activity. Rather than trying to do everything themselves, firms can therefore focus on specific phases of development; for example, a small firm developing ideas and a large one scaling them up (see the discussion on size in the previous section).

This means that open innovation is about developing in-house opportunities that have been initiated outside, but also conversely about developing outside the company some opportunities that have been initiated in-house. This can be done through, for example, spin-offs (see previous section) or through agreements to transfer IP rights to a third party (licensing). Such an outsourcing of innovations can reduce or eliminate production and distribution costs and risks as well as allowing the project to reach a larger market or target new applications. Licensing can also be used when attempting to establish industry standards or to gain access to complementary technology (through cross-licensing).

Open innovation can therefore provide various tactical and strategic benefits. From a tactical point of view, it can allow firms to reduce costs, risks and development time and to jointly develop bundled offers (solutions rather than products). For example, a media firm offering an e-book combined with best-sellers and an access to leading newspapers.

From a strategic point of view, open innovation can generate synergies and a critical mass of capabilities that allow firms to achieve economies of scale and global reach. It can also provide the firm with a privileged access to key or complementary resources and insights.

But working with others, in particular around innovations, can conflict with traditional corporate and competitive ways of managing, based on exclusivity, core competencies or proprietary information and skills. Open innovation also require firms to have a clear vision of the unique capabilities and core competences they want to develop and maintain in house, and conversely of the ones they are ready to share or co-develop.

Firms must therefore consider how best to develop open innovation capabilities in line with their innovation strategy, regarding the identification, evaluation and implementation of partnership opportunities.

Identifying partnership opportunities can be achieved through scouting, which relates to the ability to search actively for innovation opportunities among partners, competitors, customers and third parties. This requires well-networked people with sufficient internal legitimacy and adequate incentives. This can also be done using professional intermediaries and social networks, off- or online.

First, identifying partnership opportunities effectively can also be facilitated by the ability to promote and project oneself as a valuable partner. Attractive partners will approach a firm only if it is perceived to be willing and capable of collaborating effectively on innovation. For example, firms such as Procter & Gamble, Google or Apple have built such a reputation as innovative organizations that other firms constantly court them in the hope of developing new partnerships.

Second, evaluating partnership opportunities effectively relates to both negotiation and uncertainty management. Firms must be able to negotiate *ex ante* and manage *ex post* the partnership's activities and deliverables, in particular property rights and the risk of contamination (loss of freedom to operate as a result of partners' claims). This can involve significant costs related to the monitoring and enforcement of the contractual terms and can conflict with institutional constraints, such as the legal status or the governance structure of one of the partners.

The negotiation and management of innovation partnerships is particularly complex because of the high level of uncertainty involved in such projects. Uncertainties can relate to complex underlying technologies, the tacit or intangible aspects of the intellectual property involved, or to the evolution of the scope of the project. Indeed, the evolution of the respective commitment of the partners or of the potential markets targeted, the creation of new or significantly modified knowledge, delays or perceptions of free riding can all hinder the negotiation and management of open innovation projects.

Finally, the management of open innovation projects can lead to significant integration, trust and communication challenges among the partners involved, taking into account their respective culture, style and/or business processes (such as their information technology platform, IP management processes, project management methodologies, incentives or rewards). The management of open innovation projects requires in particular a careful design of the underlying governance structure, in terms of equity (considering joint ventures or minority stakes) and contracts (involving strategic alliance, collaborative research, project consortia or arm's-length contracts such as a joint marketing agreement, outsourcing or licensing).

In summary, open innovation approaches provide both new opportunities and new challenges. Managers underestimating one or the other of these can face high losses and disillusion.

Key points to take away from Chapter 4

The changes required for innovation to emerge can happen only if people and teams act entrepreneurially within their organizations and environment. The development of entrepreneurial resources must therefore be managed at four levels: the individuals; the teams; the organization; and the company's ecosystem.

At individual levels, even if factors such as the presence of role models can have an influence, it is important to understand that people are not born entrepreneurs. Entrepreneurs are people who chose to engage simultaneously in the identification, assessment and implementation of new opportunities and in the engagement of people around them. They are ready to change the way things happen and act outside the (corporate) box. They choose to act in that way because of the attitudes, social norms and behavioral control they perceive, which can be influenced by the way they are managed, in particular by the presence (or absence) of employee-level innovation metrics and rewards.

At the team level, formal or informal groups of people will be innovative if they are (i) led effectively; (ii) if the right team composition, task allocation and team dynamics are in place; and (iii) if they are faced with the right level of stress. In particular, diversity and stress can have a positive influence on innovativeness but can also hinder it, and both need to be carefully managed. Providing extra team members and slack or free time is not always a good thing *per se*.

Moreover, dedicated innovation team leaders should be trained and retained, and innovation team members should be selected based on the skills, background and attitude they bring to the team rather than because they happen to be available at the time of the launch of the project. Innovation project staffing is probably one of the situations where there is the highest discrepancy between how much managers recognize (*ex post*) that it is a key success factor and how they fail (*ex ante*) to manage it carefully.

At organizational level, firms must first define what they mean by being innovative in terms of input, processes or output, in a way that is aligned with their strategic objectives. In particular, they need to manage the balance between adopting a more innovative mindset across the organization (culture) and developing innovative businesses within the organization (structures).

The innovative culture or climate can be influenced, positively and negatively, by various formal and informal organizational factors, and in particular by the entrepreneurial orientation of the organization's leadership. This is in particular the case with large firms, which must become "ambidextrous", leveraging the benefits and limiting the drawbacks that their scale and complexity bring. Each firm must implement the balance of discipline and flexibility that fits its strategy, in particular regarding exploitation and/or exploration objectives, and carefully manage the resulting tensions.

Firms can also create dedicated corporate venturing structures, either inside their organization or at arm's length, where new business opportunities can be captured. These structures require, however, careful governance, in particular managing how much they operate within or outside the organization's social norms, systems and processes. Like "positive parasites", corporate ventures must live and grow within or close to the "organizational body" but without destroying it or being rejected.

Finally, the way firms can or cannot manage innovation is influenced by the ecosystem in which they operate. This ecosystem includes the regulatory and institutional arrangements in place, the material, intellectual and financial resources available locally, and the systems and processes put in place by public authorities to support innovations. This ecosystem also includes the networks of innovation partners with whom the firm can collaborate to decrease costs and risks, save time, access capabilities and markets, or build standards.

In particular, firms should proactively leverage the opportunities raised by their innovation ecosystem, by implementing open innovation processes aimed at more and better innovation with others while preserving their key resources and core competences.

Further reading

Amabile, T. M., Conti, R., Coon, H. and Lazenby, J. (1996), "Assessing the Work Environment for Creativity", *Academy of Management Journal*, Vol. 39, No. 5, pp. 1154–1184.

Amabile, T. M. Hadley, C.N. and Kramer, S.J. (2002), "Creativity under the Gun", *Harvard Business Review*, August.

Anderson, N. and King, N. (1991), "Managing Innovation in Organizations", *Leadership and Organization Development Journal*, Vol. 12, pp. 17–21.

Birkinshaw, J. and Hill, S. A. (2003), "Corporate Venturing Performance: An Investigation into the Applicability of Venture Capital Models", in *Academy of Management 2003 Best Paper Proceedings*, Glassboro, NJ, Academy of Management, pp. B1–B6.

Chandler, A. D. (1962), *Strategy and Structure*, reissued 1982, Cambridge, MA, MIT Press.

Chesbrough, H., Vanhaverbeke, W. and West, J. (2006), *Open Innovation: Researching a New Paradigm*, Oxford University Press.

Chesbrough, H. (2006), *Open Business Models*, Boston, MA, Harvard Business School Press.

De Long, D. W. and Fahey, L. (2000), "Diagnosing Cultural Barriers to Knowledge Management", *The Academy of Management Executive*, Vol. 14, pp. 113–127.

Dutta, S. and Caulkin, S. (2007), "The World's Top Innovator", *World Business,* January/February, pp. 26–37.

Gulati, R., Nohria, N. and Zaheer, A. (2000), "Strategic Networks", *Strategic Management Journal,* Vol. 21, pp. 203–215.

Hill, S. A. and Birkinshaw, J. (2008), "Strategy–Organization Configurations in Corporate Venture Units: Impact on Performance and Survival", *Journal of Business Venturing*, Vol. 23, pp. 423–444.

King, N. and Anderson, N. (1995), *Innovation and Change in Organizations*, London: Routledge.

Maula, M. V. J. (2007), "Corporate Venture Capital as a Strategic Tool for Corporations", in H. Landström (ed.), *Handbook of Research on Venture Capital*, Cheltenham, Edward Elgar, pp. 371–392.

Mintzberg, H. (1979), *The Structuring of Organizations*, Englewood Cliffs, NJ, Prentice-Hall.

Morris, M. H., Shaker, A. Zahra, S.A. and Schindehutte, M. (2000), "Understanding Factors That Trigger Entrepreneurial Behavior in Established Companies", in G. Libecap (ed.), *Entrepreneurship and Economic Growth in the American Economy (Advances in the Study of Entrepreneurship, Innovation & Economic Growth, Volume 12)*, Bradford: Emerald Group Publishing Limited, pp. 133–159.

Porter, M. E. (2000), "Location, Competition and Economic Development: Local Clusters in a Global Economy", *Economic Development Quarterly*, Vol. 14, pp. 15–34.

Quinn, J. B. (2000), "Outsourcing Innovation, the New Engine of Growth", *Sloan Management Review*, Vol. 41, pp. 13–28.

Screiber, N. and Chakravarthy, B. (2007), "Leading Paradoxically", *European Business Forum*, Vol. 30, pp. 29–33.

Shook, C. L., Priem, R. L. and McGee, J. E. (2003), "Venture Creation and the Enterprising Individual: A Review and Synthesis", *Journal of Management*, Vol. 29, pp. 379–399.

Teece, D. (2009), *Dynamic Capabilities and Strategic Management*, Oxford University Press.

West, M. A. (2002), "Sparkling Fountains or Stagnant Ponds", *Applied Psychology: An International Review*, Vol. 5, pp. 355–524.

CHAPTER 5

Sourcing: finding the gold nuggets

> **Short case: identifying opportunities at Materials International[1]**
>
> Materials International is an industrial firm supplying the construction and automotive industries. It has a long history of introducing new versions of its core products, based on world-class in-house R&D developments and a strong intellectual property portfolio. Faced with both a crisis affecting its main customers and the emergence of new competitors from Asia, it decided a few years ago to focus its strategies on innovation. It quickly realized that in-house R&D was not enough as a source of innovations and decided to launch a wide range of bottom-up and top-down initiatives aimed at identifying innovation opportunities. The main initiatives are described below.
>
> The first initiative was to increase the level of R&D spending in spite of the economic crisis. This increase in resources was combined with a refocus of the R&D portfolio around a small number of key industry challenges and opportunities, linked, for example, to the control of greenhouse gases or the development of new materials used for renewable energy production. The head of R&D was also promoted to Vice President (VP) responsible for innovation, working closely with the CEO.
>
> $\rightarrow$

The second initiative was to set-up a dedicated team, reporting to the VP Innovation and the CEO. The mission of this new business team was to identify and develop three to five high-potential opportunities to introduce radically new value propositions and business models. This team has a high degree of freedom, including exploring opportunities to restructure the existing value chain or to develop strategic partnerships and enter completely new markets.

The third initiative was to set up throughout the business units formal innovation development centers, consisting of local managers representing all the main functions, meeting regularly and reporting to the VP Innovation. The mission of these centers is to identify and develop local opportunities to improve existing products and processes, other than the technical improvements already developed by R&D. These innovations included opportunities such as a reorganization of the supply chain or the introduction of new product designs based on existing technologies.

The fourth initiative was the launch of a company-wide project to stimulate the involvement of all employees in the identification of innovation opportunities, with strong support from corporate and local managers. All employees have been provided with access to training and to a new online opportunity-sharing tool. Local innovation gatekeepers have been appointed to support and guide them, and individual reward systems have been set up in each country where the company has facilities.

These initiatives have already led to the successful launch of a new product range, a completely new business model (based on interactive advertising), and significant savings throughout the organization. The firm has also improved its reputation significantly as an innovative company, in particular in the eyes of high-potential recruits. Further initiatives are planned, focusing in particular on the proactive identification, evaluation and implementation of partnership opportunities.

Having defined an innovation strategy and put in place the entrepreneurial organization to support it, the next key challenge is to identify potential innovation opportunities that could allow for the paving of the way to that strategy. A brief reminder here that, as discussed in Chapter 1, what matters is not to invent or generate ideas but rather to identify innovation opportunities, including some that might have been initiated by others.

The first step is to identify where to look; that is, what are the main sources of innovation that the organization can tap. These sources include R&D activities but go far beyond them. The second and third steps are to define how to harvest this flow of opportunities in a proactive and systematic way, both internally and externally. We shall discuss these three aspects in the three sections of this chapter.

Sources of innovation

As discussed in Chapter 1, innovations do not arrive out of the blue. Innovations are about new combinations occurring in specific contexts. Understanding where such new combinations can emerge and in what context is therefore critical. Here, we shall first discuss the classic source of technology innovation – research and development (R&D) activities –then highlight how sources of innovations go far beyond R&D, and finally discuss how to integrate these various sources.

Sources of technology innovations: R&D

Since the eve of the Industrial Revolution, firms have been exploiting scientific discoveries in order to develop new products and find better ways of delivering them. To stimulate scientific research and its application in specific fields and in an exclusive way, many firms have developed in-house research and development activities. As a consequence, R&D spending has long been used as a proxy for the intensity of the innovation activities of an organization or an industry.

The R&D intensity of firms varies widely across sectors,[2] from being marginal (less than 5 percent) in the service sectors and in some heavy industries such as pulp and paper or oil, to more than 20 percent in the so-called high-tech sectors such as electronics, computer software or healthcare. Hence firms in the automobile, pharmaceutical and information technology sectors spend billions of their own money each year on hiring and training experts in order to develop proprietary scientific research and development.

Developing an innovative R&D capability supposes, on the one hand, the development of an R&D strategy, to define where and how much time to spend looking for opportunities, and on the other, to set up an R&D organization to define where to place resources and how to integrate them.

The R&D strategy must first be defined according to the environment, resources and purpose of the organization. It must in particular support the development of the core competencies of a firm. Indeed, to quote one of the most famous papers on management: "The real sources of advantage are to be found in management's ability to consolidate corporate-wide technologies and production skills into competencies that empower individual businesses to adapt quickly to changing opportunities."[3]

Once the core areas or technology platforms a firm wants to focus on have been defined, it must decide how much it wants to invest across business cycles. In particular, firms facing decreasing sales or profit must decide how to manage their R&D expenditure effectively, taking into account both short-term profitability (in other words, try to spend less) and long-term perspectives (in other words, try to spend more or better).

Deciding what the right level of R&D spending is can be done through a bottom-up approach, based on the probability of technological and commercial success among the opportunities identified, and on the corresponding expected project development costs, timing, markets and profits. This can also be done through a top-down approach, based on the benchmarking of peers and/ or on strategic decisions about the areas the organizations wants to learn from and build its future on. The first approach fits well with an exploitation strategy focused on incremental innovations, while the second is more in line with an exploration strategy where more radical innovations are expected.

Having decided how much of its resources to invest in R&D and where, the firm's next step is to design and implement the R&D organization that will best support those choices. In particular, in the context of multibusiness and/or multinational firms, R&D departments are no longer merely corporate black boxes attached to headquarters and with limited monitoring of their activities and output. How and by how much they are centralized or distributed has become a key decision in terms of business process, corporate structure and geographical scope.

First, as R&D activities and performance become integral parts of their business's processes, firms must decide how much R&D activities are managed through dedicated formal structures, gaining focus, accountability and discipline, or through relatively informal networks, gaining flexibility, market orientation and agility (see the discussion on ambidextrous organizations in Chapter 4).

Second, as the corporate structure combines businesses active in different competitive environments, R&D activities must be balanced carefully between corporate (typically long-term) and divisional (more medium or short-term) ones, both in terms of funding (who decides on and who pays for the projects) and of absorption (who integrates and who implements the opportunities developed).

Third, firms active across various regions must define the geographic scope of their R&D. This should take into account local market characteristics but also where the best talents and expertise can be found, which might be far away from corporate headquarters. In the latter case, R&D should be balanced between, on the one hand, global or regional structures connected with a wide array of production units, potential recruits and customers, and on the other more local structures fostering local connections and close collaboration. This is particularly critical considering the emergence of a highly skilled workforce, sophisticated customers and very competitive firms in emerging countries such as the BRIC countries (Brazil, Russia, India and China).

As a consequence, firms have adopted a wide array of R&D organizational structures, from networks of specialized regional centers with local market reach and focused technology scope, to centralized structures with broad technology scope and global market reach. Others have developed transverse approaches based either on technology centers of excellence, with a focused technology scope and global market reach, or on regional centers with a broad technology scope and local market reach. Which style is chosen should be based on what best fits the R&D and overall strategy of the firm, considering its specific resource, environment and purpose, and in particular balancing local flexibility with global efficiency.

But while deciding effectively how much, where and how to develop R&D capabilities might be a necessary condition to compete in many industries, there is ample evidence that it is not a sufficient condition for success. Other sources of innovations must be leveraged in order to compete effectively, which will be discussed in the next section.

Sources of business innovations: beyond R&D

While the level of R&D spending remains a popular proxy for innovativeness among economists, policy-makers and some managers, there is ample evidence that the amount a firm spends on R&D is a poor indicator of how much it innovates and an even worse one of how *well* it innovates.

Indeed, there are many very innovative firms that do not invest significantly in R&D. In fact, from a quantitative point of view, R&D is not even the most important innovation input – it is the level of investment in capital equipment related to the introduction of new products.[4]

Furthermore, across sectors, higher R&D spending does not ensure higher performance, be it in terms of growth, profitability or shareholder returns. Multiple studies have found no discernible statistical relationship between the level of R&D spending and various measures of business success.

As an example, Dell revolutionized the personal computer industry while spending much less (in absolute and relative terms) than its main competitors. Similarly, large pharmaceutical firms dwarf small biotech firms in terms of R&D expenditure but tend to be much less effective, taking into account the number of drugs that reach regulatory approval. Finally, the US firm that had the highest level of R&D spending during the twenty-five years up to 2010 is... General Motors (GM), which was on the brink of bankruptcy in 2010 and had to be bailed out by the government.

One potential explanation for this apparent disconnect between R&D spending and business performance is related to poor R&D management, as R&D investments can be wasted on unsuccessful or irrelevant projects. But thirty years of improvements in the way that R&D is managed across firms has not resolved this apparent paradox, and it would be very bold to assume that R&D is still poorly managed in the majority of firms.

An alternative explanation is that there are many other sources of innovation beyond R&D, either in other parts of the organization or outside it. Firms can therefore be successful by leveraging those other sources. Conversely, firms that fail to do so might lose their competitive edge even if they have strong R&D capabilities.

There are indeed various types of discontinuities that can trigger the emergence of an innovation, hence various directions in which to look for innovation opportunities. These are detailed below and can relate to challenges, evolutions or revolutions.

The first category of sources of innovation relates to *challenges* to received wisdom or existing assumptions. Such challenges might result from the unexpected success or failure of a project or experiment. From whisky to X-rays, history is indeed full of accidental discoveries.

Such challenges can also be caused by an unexpected market reaction, when an offer provides different benefits than those for which it was designed. As an example, several drugs having positive side-effects became successes based on the side-effects rather than on their initial purpose (see the short case in Chapter 7). Another example is the first interactive cell-phone services, which were aimed at affluent businessmen but were in fact adopted by teenagers and gamers.

In this situation, the issue for the organization is to recognize in such an unexpected event an opportunity to be explored, rather than a deviation from predefined plans or objectives. This is particularly difficult in fragmented and objective-driven organizations, acting like silos. In such structures, an unexpected event in one department often fails to be shared and exploited across other parts of the business, and it is eventually disregarded because it does not fit the specific scope and objectives of that department, or because other entities are not aware of it, or receptive but "mind their own business".

Challenges to the received wisdom or existing assumptions can also emerge through positive deviants, individuals or organizations questioning prevailing orthodoxies and exploiting inadequacies in underlying processes that are taken for granted. These include traditions, routines or we-have-always-done-it-this-way situations. This is the case, for example, in the wine industry, where new producers have challenged the "Grand Cru" traditions and methods, or in the accommodation sector, where new entrants have rejected the one-star to five-star classification system. Here, the issue for the organization is, in particular, its ability to tolerate and give some space to such positive deviants, bending the rules and challenging existing business processes and bureaucracies (see the discussion on corporate entrepreneurs and innovative teams in Chapter 4).

The second category of sources of innovation relates to *evolutions* in the environment that the firm faces, creating both new threats and new opportunities. These evolutions may relate to changes in the industry and the market structure that shake the incumbent's business model. This includes, for example, deregulation or disintermediation processes in the retail and banking sectors.

Those evolutions may also relate to new political environments leading to new policies or new governance approaches; for example, involving nationalization or support for national champions, such as in the energy or steel sectors.

The main issues for the organization in this situation are to sense when and where political grounds are shifting, to identify and influence whenever possible the appropriate stakeholders (through lobbying, standard-setting and advisory committees among other methods) and finally to put in place the right change management mechanisms.

Evolutions in the environment the firm faces can also relate to its existing and potential markets. The nature and structure of these markets can evolve – for macroeconomic or demographic reasons such as migration, aging or urbanization.

The needs and expectations of existing and potential customers can also evolve, as new generations and new moods emerge, and as mental models and values change – in this case perhaps with respect to environmental values, new family structures and changing religious or cultural beliefs. This affects, for example, innovations in the financial service industry, where new investment products are matched to specific communities, personal situations or risk profiles.

Finally, the third category of sources of innovation relates to *revolutions* in processes and technologies. Those can relate to new necessities, as existing approaches and processes become unsustainable from an environmental, social or financial point of view. Examples of this would be in the transportation sector or in energy-intensive industries.

Such revolutions can also relate to opportunities, as new technologies and new areas of knowledge appear, which can be integrated with new business models – for example, relating to new materials, new drugs or new ways to gather and deal with information.

Hence a firm focusing exclusively on R&D as a source of innovation would probably miss threats and opportunities emerging from the first two categories outlined above (challenges and evolutions) while only partially leveraging the third category (revolutions).

Developing the skill of identifying innovation opportunities therefore requires firms to combine, leverage and balance various internal and external sources of innovations proactively, beyond R&D. This aspect is described next.

Combining internal and external sources of innovation

An innovation opportunity, whether triggered by a challenge, an evolution or a revolution (see above) may be initiated either inside or outside the organization.

In the first case, a new solution or a new technology is conceived within the organization, either by finding a new way of using an existing piece of technology, by exploiting new offers from suppliers, or by inventing something new. This new solution is then "pushed" towards the firm's market and environment, to successfully implement or commercialize the resulting innovation.

This traditional way of developing an innovation, derived from the way that R&D activities used to be managed, can still be very original and creative. Indeed, it has led in the past to breakthrough innovations such as the laser, the disposable shaver and light emitting diodes (LED).

However, such an approach can also lead to a great deal of wasted effort and resources, if products that do not meet any relevant or significant needs are developed. It is therefore in general better suited when the objective of the firm is to identify *radical* innovation opportunities, where firms attempt to create or tap into new needs.

Let us stress that, in this first case, the new solution or technology can be initiated through various technology trajectories within the firm, and not only by its internal R&D department. Indeed, the emergence of new technology solutions can be driven by the development of complex production systems and infrastructures or by specialized suppliers active in niche markets, such as healthcare or heavy industry. It can also be driven by technical changes among suppliers, in particular suppliers of information technology systems, such as in the financial or retail sectors, where such new technologies have led to radical process innovations such as online banking or automated trading.

As a consequence, such an approach should be supported by the development of technology intelligence capabilities beyond one's own R&D, product roadmap or experience curve. This includes tracking publications and patents (both in terms of frequency and content) and attending open technical meetings. It also includes benchmarking and reverse engineering, acquiring and licensing in technologies, and hiring or interviewing competitor's employees.

But innovations do not only emerge from inside the organization. The starting point of an innovation can also be a new need, or a new idea about what could be offered. In the latter case, the innovation opportunity is "pulled" by the organization, either through trendspotting, market research or needs analysis techniques such as quality function deployment. In this case the challenge for the firm is to identify new market needs better and/or faster than others and to implement effectively relevant technological solutions.

This approach can lead to very efficient and reactive project implementation, as only what is perceived as being needed by users or customers is provided. However, it can also lead to short-sighted and/or ineffective approaches. Market research often either fails to reveal latent needs or only reveals information available to all competitors, which generates limited opportunities to develop a differentiated offer.

Market research often struggles with innovations, because most people have difficulty in talking about what they have not yet experienced. They tend therefore to focus on incremental or obvious innovation opportunities. Past market research misses include major innovations such as the personal computer, cell phones or the Walkman. This approach is therefore in general better suited when the objective of the firm is to identify continuously *incremental* innovation opportunities.

One way to implement such a "pull" approach is to rely on lead users. These are users who tend to recognize their needs before mainstream customers (they are "awake") and who expect to benefit from and/or contribute to the development of new solutions (they are "interested"). Let us stress, however, that relying on these lead users can reduce a firm's control of the process and expose it to some degree of customer opportunism. It can also generate interactions that are complex to manage, in particular regarding the leakage of information and the loss of property rights.

Having reviewed some key aspects of both "push" and "pull" approaches, three things must be highlighted regarding the use of these. First, both approaches (pushing/creating waste and pulling/efficient short sight) have their benefits and shortcomings. Relying exclusively or excessively on one of them is probably a recipe for failure. Hence each firm has to identify the balance of "pull" and "push" approaches that fits its strategy, and in particular whether its resources, environment and purpose support the identification of incremental or radical innovation opportunities.

Second, combining and integrating "push" and "pull" approaches often requires mixing people and teams that have very different cultures, values and mental models. It can therefore be productive but also very tricky (see the discussion about diversity in innovative teams in Chapter 4). The archetypal engineer, focused on building, reinventing and improving features and on perfection rather than speed, will often conflict with the archetypal marketer, focused on conviction rather than feasibility, and on selling, reusing or repackaging good enough things that can provide tradable benefits.

Third, as discussed in Chapter 1, innovations are not just ideas that pop up instantaneously out of the blue. Organizations might in some cases want to look for early innovation opportunities or new concepts. They are often relatively easy to find and cheap to acquire; for example, by scanning inventors and patents directly or through specialized intermediaries. But organizations should also look out for more mature innovation opportunities, those that have already been developed and tested – for example through replication, licensing from or outright acquisition of existing businesses. These can be more difficult to find and more expensive to acquire, but they can also significantly decrease the risk of failure and reduce the time to market.

In other words, organizations must look for a wide range of innovation opportunities, scanning both within and outside their boundaries (not only in R&D), and both new concepts and more mature opportunities.

Again, whether the innovation was invented or initiated by the organization itself is not what matters from an innovation management point of view. In particular, raw ideas that have been discovered within the firm which must still be tested and developed are often not the most efficient or most profitable way to identify innovation opportunities. Many firms, having implemented simple innovation portals or equivalent idea-box mechanisms, have indeed struggled with such approaches.

Having discussed the various sources of innovation as well as their integration, we shall review in the next two sections the specific mechanisms and processes organizations can put in place to harvest internal and external sources of innovation opportunities.

Harvesting corporate knowledge (internal sources)

As discussed above, the first source of innovation opportunities for an organization is the ideas and opportunities it identifies and pushes within its ranks. Proactively harvesting these sources of innovation stimulates the creation of new knowledge. This means developing more effective search behaviors, to manage effectively the knowledge created throughout the organization, and finally to protect the knowledge created and build competitive advantages from IP rights. These three aspects will be addressed below.

Creating new knowledge: search behaviors

Search behaviors relate to the capability of an organization to acquire and process new information on a continuous basis. While it tends to happen all the time across most organizations, it can actually be structured along three dimensions: "problematic", "slack" and "institutional" search behaviors.

"Problematic" search behaviors arise when an issue emerges that challenges current practice. It refers to the process through which managers react when organizational performance falls below historical levels and/or below social aspirations. It can be triggered by internal or external benchmarks, for example, or by reverse-engineering competitors' products and practices in order to discover gaps in performance.

One important issue when identifying innovation opportunities through problematic search behaviors is defining the intensity of change involved. When the performance or results achieved is deemed to be unacceptable, an organization can first try to adjust its strategy or action plan through incremental innovations, aimed at better exploiting the current situation. In this case, minimal change and questioning is involved.

In other cases, however, adjustment alone is not enough: the organization needs to reconsider and challenge its core values, develop radical innovations and explore new ways to run its business.

In other words, in problematic search, the issue is to decide when evolution is sufficient and when revolution is needed.

As an example, faced with the growth of media contents available online for free and the resulting decreasing sales, incumbent media firms can work to evolve and create an improved product (such as high-definition (HD) or 3D movies). But they can also try to revolutionize their industry, completely change their business model and aggressively move online through for example video-on-demand offer, social networks or file-sharing services. In this second case higher long-term profits might be at stake but also higher uncertainty.

In contrast to "problematic" search, "slack" search behaviors may emerge when the level of organizational stress is not too high (see Chapter 4) and when sufficient resources (time, money and management attention) are available to be invested in new solutions to improve performance. As it is not directly issue-driven, such an approach may induce actions in areas that do not directly affect performance – for example, by mobilizing the tacit knowledge of the organization (see the next section). Slack search behaviors can also be stimulated through the use of creativity techniques, such as brainstorming, using metaphors and word associations, or adverse thinking.

The third type of search behavior, "institutional" search, is the type carried out by specific organizational units such as corporate R&D (see the previous section) or marketing research departments. For these departments, the core activity *is* to identify new opportunities. They are often not tightly linked to business performance, either in terms of purpose or timing.

Institutional search can also happen through dedicated opportunity aggregation processes such as open idea portals or business plan competitions. In this case, the organization puts in place dedicated processes allowing any employee (or some specific subgroups) to develop and submit innovative ideas. While these tools have been implemented in many organizations and regions, they require careful design and implementation in terms of submission gathering, coaching and feedback as well as project ownerships.

It must be stressed that such opportunity aggregation processes often have to rely on effective facilitators who act as contact points, gatekeepers or mediators. Those are people within the organization who can, formally or informally, act as repositories of knowledge, know who else possesses useful knowledge, have the skills required to make connections, and can therefore act as legitimate go-betweens for different parts of an organization. They can, for example, foster the creation of communities of practice across and outside organizational boundaries.

Successfully implementing such an approach therefore goes far beyond setting up the underlying web systems and processes involved. In particular, understanding what is not relevant as an opportunity and what is very new but still relevant for the firm is often not obvious to most employees. They risk therefore either clogging the process with irrelevant submissions, or on the contrary suggest only marginal improvements.

Similarly, being able to synthesize, submit and present a new idea (a short business plan) in a meaningful way is not an innate skill: it requires training and support. As a consequence, these opportunity aggregation processes tend to have a limited direct effect on corporate performance. They can, however, have strong positive indirect effects on the internal atmosphere, boosting entrepreneurship, networking and external reputation.

Having reviewed the three types of approach that firms can implement in their search for new knowledge, in the next section we shall discuss the issue of managing that knowledge effectively.

Managing new knowledge

Organizational learning is fostered in organizations where individuals not only have the capacity (as individuals or as a community) to invent new behaviors, but also have the ability to move around, and where there is an established process for the transmission of a skill from the individual to the community as a whole.[5] As discussed in Part I, innovation is about new combinations more than new ideas. For example, throughout history, populations living in large landmasses where travel and exchanges are facilitated have been more innovative than small, isolated populations.[6]

The challenge for an organization is therefore to stimulate such transfer, combination and appropriation of knowledge. The purpose of such knowledge management for the organization should be not only to capture and combine explicit factual information in a systematic way but also to leverage the tacit information and knowledge that exists internally. In particular, it is about better leveraging the experience and learning of individual employees in order to advance the organization's mission.

The knowledge or intellectual capital of an organization can take various forms. It includes the knowledge and know-how of its employees as well as their competencies (skills, education, experience and training). It also includes structural capital such as intellectual property (IP) (see the next section) and trade secrets, licenses and franchises as well as in-house databases, software, work documents and publications. Finally, it also includes the market capital of a firm, embedded in its collaborations, partnerships and networks through lists of customers and suppliers, product certifications, brand and goodwill.

A key issue regarding knowledge is the difference between explicit and tacit knowledge, both of which coexist in organizations. Explicit knowledge relates to formal processes, rules, tools and methods. It is about structured, codified and declarative facts ("I know that...") which can be documented in an unambiguous and generic way. It can be validated by experts and transferred. Explicit knowledge fits well with a mechanistic view of management (see Chapter 4).

On the other hand, tacit knowledge concerns the fact that, as human beings, we know more than we are able to explain, be it in terms of knowing how, when, where or why things should be done in a specific way. It relates to contextualized and procedural knowledge ("I know how to do things") which is by nature subjective and idiosyncratic. Such knowledge is personal and can be conscious or not. It is based on experience (rather than expertise) and rooted in action.

This distinction is important because the nature of knowledge will condition the absorptive capacity of the organization, which is its ability to recognize, assimilate and utilize new knowledge. The absorptive capacity of an organization is driven by its prior investments in learning and the resulting accumulated knowledge and experience, by the presence of complementary technologies and accessible partners, and by its readiness to learn new things.

The way such absorption can happen and/or can be stimulated will depend on the nature of the knowledge involved. Tacit knowledge can be absorbed through networking, imitation and teamwork (for example, through role models) or other "person to person" channels. Explicit knowledge tends to be transferred from individuals to the organization through formal decision-making processes such as analyses or computation.

Organizations can also engage in externalization, when individual tacit knowledge is codified into organizational explicit knowledge through formal reporting, training, prototyping or other "person to document" channels. Indeed, if an organization relies too much on uncodified tacit knowledge, it can become unmanageable, as nobody dares to touch or change anything. Conversely, internalization happens when individual explicit knowledge is tacitly appropriated by the organization, through, for example, on-the-job learning and experimentation.

Effective knowledge management allows an organization to better share and combine such various types of knowledge among its members and to leverage more effectively the knowledge they each generate. In particular, it helps the organization to minimize repetition (reinventing the wheel) and the loss of critical knowledge as a result of employee turnover. It is unfortunately often only when a key person leaves the firm or when distant departments meet that the importance of knowledge management is understood and proactive processes are put in place.

Knowledge management typically relies on information technology systems and processes (such as repositories or tools) but also on adequate human resources and IP management. In particular, employees must be trained and motivated to generate and share knowledge, and the tools put in place must meet their needs in terms of availability or ergonomics. Again, it is about people, not just systems and processes.

Indeed, knowledge is a good that has peculiar characteristics and therefore requires specific approaches. Knowledge can be idiosyncratic, limiting its applicability or relevance throughout the organization. This is particularly the case for organizations formed through mergers and acquisitions, which must define how much common knowledge can be shared and applied across various units with a range of history and experience. Failure to do so is at the root of much value-destroying empire building.

Knowledge also tends to be generated in vast quantities, beyond the amount that any human brain can master, leading organizations to specialize more and more in order to cope with the bounded rationality of their employees. Organizations must therefore in particular define which knowledge is deemed to be critical, based on its complexity, scarcity and relevance. This implies a definition of the critical domains where knowledge should be capitalized, who are the key people from whom such knowledge can be transferred, and how better to share and develop such knowledge.

But knowledge is also a public good, which means that it is non-rivalrous and non-excludable. To put it simply, it can easily be copied. Building competitive advantages through knowledge therefore requires organizations to define, claim and protect intellectual property rights, which are discussed below.

Protecting corporate knowledge

To motivate economic actors to invest in the exploration of innovation opportunities and avoid quickly being copied by free-riders, public authorities across the world have created regulations providing inventors with some degree of exclusivity. Such intellectual property (IP) rights are legal rights over the creative ideas of an individual's mind, whether technical or non-technical. They relate to exclusive rights over the use of something new for a certain period of time. This can relate to documents and presentations (copyrights), databases, software, designs (registered designs), to know-how or to the protection of reputations (trademarks).

Let us stress that while copyright is automatic (there is no need to take a specific action to benefit from it), other IP rights require the creator to take specific actions to claim his or her rights.

Regarding the protection of inventions, obtaining a patent provides the exclusive right, granted by a government, to exclude others from making, using, marketing or selling the invention for a certain period of time (usually twenty years). Hence patents provide temporary private benefit aiming at the long-term public interest. They therefore provide incentives to innovate and can facilitate the diffusion of technology, foster the creation of firms and facilitate technology markets.

However, patents can also be used anti-competitively, create monopoly distortions and block follow-on innovations.[7] They play, for example, a very important role in the information technology sector, where the most prolific "patents machines" are to be found, and in the pharmaceutical sector, where incumbents must disclose for regulatory and safety reasons their formulas and therefore need protection from copycat or generic products.

Patents apply in principle on inventions or technical creations, not discoveries, though in some countries discoveries such as genes can be patented. These inventions must fulfill certain conditions (novelty, inventive step, useful and applicable at industrial level and so on) and the patent is granted by a government for a limited time and a limited extent of protection. Let us stress that patents provide a right to exclude others, not a right to make, use or sell. It is not because a firm has a patent on something that it has freedom to exploit it.

To gain patent protection an inventor must disclose or describe his/her invention and justify its novelty (it is not part of the state of the art), inventivity (it is not trivial or obvious for a skilled person) and applicability (an example of actual implementation can be presented). The inventor must also specify the extent of protection requested, both in terms of geographies covered and types of applications. These "claims" are validated or invalidated by the relevant patent office and can be contested in courts.

What is important to stress is that, here again, patents are not good things *per se* and the number of patents is a very dangerous criteria for assessing the performance of a business (or region). Many patents are applied for but never used, and a patent is in no way a validation of the quality or the value of an invention (only of its novelty, inventiveness and applicability – see above).

Patents can bring important direct and indirect benefits to the firms claiming them, but they also have significant direct and indirect drawbacks. These are discussed below.

Patents can provide significant direct benefits to their owner. First, by definition, they are a way of acquiring exclusive rights over technologies and to control their development. Second, they allow an organization to convert inventions into valuable tangible assets that can be accounted for, rented (licensed) or sold. It is indeed much easier for a firm to sell or provide as collateral a patent (which is a tangible asset) than it would be for an invention alone (which is not). Finally, patents can provide their owners with an innovative and/or credible image.

From an operational point of view, patents can therefore be used to gain local exclusivity, and protect core and related or associated technologies as well as to maintain a good reputation and leadership. Firms such as Qualcomm and IBM have, for example, extracted strong revenue streams from their patent portfolio.

Patents can also provide significant indirect benefits. First, as they are published, they prevent competitors from patenting and can signal competitive intention. They can also provide their owner with a strong bargaining position *vis-à-vis* existing or potential competitors – for example, for negotiating cross-licensing. Finally, patents can be used to preempt accidental or malicious leaks of information from staff members as well as to motivate them and reward their innovativeness.

From a strategic point of view, patents can therefore be used to block the entry of competitors, keep them away from specific areas, or surround them. Patents can also be used to build strong negotiation positions and prevent others from patenting their product. Technology-intensive industrial firms such as those in the chemicals and petrochemicals sectors indeed engage in such "patent chess", where dissuasion tactics are common and where the same firms can collaborate on some technologies but compete fiercely over others.

But patenting an invention can also generate significant direct and indirect disadvantages. First, by definition, patenting supposes the disclosing of one's invention. The patented technology then becomes public knowledge and competitors can invent around the patent or create illegal copies. More generally, patenting in a field might signal that one is convinced of the competitive value of that field and therefore draw the attention of competitors.

As a consequence, firms must rely on expert lawyers in order to minimize the risks of disclosure while still retaining protection. Some firms even develop counter-intelligence measures such as patenting in multiple fields simultaneously, including in some where they have no strong interest, in order to hide their actual priorities.

Furthermore, patent protection is valuable only if the firm has both the opportunity and the means to maintain and defend its property rights. First, filing and maintenance costs can represent significant budgets for firms such as SMEs, in particular in Europe where patents have to be translated into multiple national legislations. More critically, protecting one's patent through the courts can generate huge legal and commercial costs, as litigation procedures may last years and significantly damage the litigant's reputation. Finally, a patent is worthless if it is impossible for the owner to detect and fight infringement, either because it is hidden deep inside the infringing firms' processes, difficult to prove and/or far away, or because the infringement happens in regions with weak or corrupt legal systems.

This challenge to enforce patent protection and the related costs can lead in some cases to guerrilla strategies, where firms with deep pockets threaten smaller competitors with lengthy and costly litigation procedures, or where infringements remain unpunished because the patent owner does not have the means to detect and fight them.

Patents also have indirect disadvantages. Locking in a technology can lead to slower adoption by consumers and/or complementors, who do not want to become over dependent on a single supplier. Patents can also lead to the duplication of research activities as firms develop overlapping or less effective approaches in order to avoid existing patents – for example, in the biotech sector.

To deal with these indirect disadvantages and boost their activities, some firms (in particular in the software industry) rely on open source models. In this case, property rights are designed specifically to allow a large community of users and complementors to access and contribute to the development of a technology. Rather than selling the core technology, firms can in such cases provide sales and maintenance, integration or complementary services as well as customized versions.

Given these direct and indirect disadvantages of patents, it is important for firms to consider alternative approaches to protect the competitive advantages derived from the development of new technologies. Indeed, famous inventions such as the Post-it note or the recipe for Coke have not been patented but yet have created significant value.

First, firms can choose to rely on confidentiality, trade secrets and/or the development of tacit knowledge to protect their invention and avoid disclosures. This implies in particular the careful definition of document management processes, and dealing with the potential leakage risks linked to staff turnover and poaching.

Second, firms can rely on their speed to market, as competitors do not have enough time to gain significant market share before the technology becomes obsolescent or commoditized. For example, in the electronics sector, by the time a new chip is reverse-engineered and the required manufacturing capabilities have been built, the technology is often obsolescent.

Finally, firms can develop around their technology a set of complementary manufacturing, sales or services assets that would be difficult or costly to copy.

Having discussed some of the key aspects of generating, managing and protecting corporate or *internal* knowledge, we shall discuss in the next section ways for firms to harvest *external* sources of innovations proactively.

Harvesting the environment (external sources)

There are many ways a firm can identify innovation opportunities within its environment, some of which have already been discussed. A firm can and should develop intelligence regarding the main trends affecting its environment, and in particular – but not exclusively – the development of new technologies. A firm should also be open to collaborating with other actors in its ecosystem (see Chapter 4) in order to enlarge its horizon and generate creative collisions of ideas and perspectives.

There are two other specific ways to harvest the environment proactively that will be discussed below: a way to harvest the market through value gap analysis and a way to harvest new ventures through corporate venture capital initiatives.

Harvesting the market: value gaps

As described earlier in this chapter, one way to identify innovation opportunities is to "pull" needs from the market and try to find effective ways of meeting them. One challenge in this approach is that a firm and its competitors by definition share the same market and therefore in theory target the same needs. There can therefore only be limited room for differentiation. A second challenge is, as discussed above, that traditional market research techniques tend to reveal only opportunities for incremental improvements.

One way around those challenges is to implement a value gap analysis. The purpose of a value gap analysis is to uncover hidden needs by mapping existing offers and customer expectations systematically against a set of identified features. The value gap analysis then consists of reviewing the perceived value of available and potential offers with respect to each of the features considered and identifying which of these features could be eliminated or created, and which could be reduced or raised. For example, low-cost airlines have focused on increasing features such as perceived value (through low price) and ticketing (through straightforward online offers) while decreasing features such as catering, connectivity, loyalty programs and seat space – the features focused on by most traditional airlines. Similarly, new hotel chains such as Formule 1 have improved the sleeping experience (bed quality, hygiene, quiet) *vis-à-vis* one or two-star hotels while decreasing other costly features such as architectural design or the provision of restaurants.

Hence the key to a value gap analysis is to identify – for example, through quality function deployment techniques – features that matter to customers but which traditional offers fail to deliver, and at the same time be ready to decrease other features on which competitors focus heavily. This allows a firm to rebalance its product or service and maintain a good overall value and/or to extend the value offered and create new sources of differentiation.

Examples of value gaps that could be exploited include:

- Offering new features such as traceability, recyclability, energy savings or ethical guarantees with traditional products. For example, automobile windshields now offer new features (beyond high strength, low weight and low cost) such as heat and fire protection, sunlight filtering or color shades, ease of cleaning/wiping or steam removal.
- Focusing on other aspects of the value proposition than the product itself (such as packaging, ease-of-use, robustness, delivery or total cost of ownership).
- Focusing on quality, simplicity or reliability in industries where the focus is mainly on technological sophistication or brand and market share development, such as for some software or automobile firms (see the discussion on disruptive innovation in Chapter 1).
- Exploiting gaps between customers' expectations and/or perceptions and industry internal performance indicators. For example, some utilities or transportation companies will focus on average quality of service while customers will remember

one day of service breakdown – one letter lost, one blackout or one long delay – much more than the 99 days of good service.

It is important to stress that value gap analysis can allow firms to de-commoditize their product or service and create significant value but it is also quite risky given the uncertainties and resistance to change faced by the radical innovations it often involves. Firms must therefore consider whether such high risk/ high potential approaches fit their strategy.

Having discussed value gap analysis as a proactive way of identifying pull innovation opportunities in its markets, we shall in the next section discuss corporate venture capital as a proactive way to identify innovation opportunities through open innovation with new ventures.

Harvesting new ventures: corporate venture capital

As discussed in the Chapter 4, small firms are not only much more numerous than large ones, but can also sometimes be better positioned to explore innovation opportunities, as they often have a lower bureaucratic inertia, a higher individual incentive to innovate and fewer constraining strategic commitments. Many large firms have therefore tried to create such favorable settings through the development of internal ventures (see Chapter 4).

Alternatively, corporate venture capital (CVC) is a way for a firm to benefit from the innovativeness of *external* ventures, through the direct investment of corporate funds in start-up companies. CVC refers therefore to initiatives at various levels of a corporation where investments are made, directly or through an intermediary, in independent companies. In other words, it is a way for large firms to be involved in the innovation process of small ones through financial investment.

CVC can allow the parent company to implement specific approaches for each autonomous venture, build new competencies and foster a more entrepreneurial culture. Indeed, CVC initiatives have been shown to have a positive impact on the patenting rate of corporations,[8] particularly when the activities of the parent company and the venture are closely related. CVC also has a positive impact in terms of recognizing technological discontinuities.[9]

A CVC initiative can be operated as a self-managed fund; this is a subsidiary where the parent firm is the sole fund provider, with tight cost control and a strong influence on decision-making. A CVC initiative can also operate through pooled finance, functioning as an independent fund where the main or most of the limited partners are corporations. In the latter case the parent corporation has in general a veto right on the investment decisions of the general partner managing the fund.

While CVC initiatives can provide independent ventures not only with cash but also giving favorable access to corporate assets, it can conversely provide the parent company with strategic benefits in terms of leveraging, option building and learning. These will be discussed below.

First, investing in new ventures can allow the parent company to better leverage its own technologies and platforms and/or its other corporate assets. Investments in new ventures can generate exposure and access to emerging complementary or disrupting technologies, support and influence the development of new applications, and stimulate and shape the demand for the parent company's technologies.

External ventures can also add their products to corporate distribution channels, utilize excess capacity and provide attractive professional development opportunities for corporate staff. They can also be used to put pressure on and/or benchmark internal suppliers.

Second, investing in new ventures can allow the parent company to build options for future business development, as initial investments can lead to outright acquisitions with significant strategic value. Investing in start-ups also allows the parent company to build contacts and networks with entrepreneurs, scientists and other investors that could lead to further innovation opportunities.

Finally, investing in new ventures can allow the parent company to learn in a fast-track way about new markets, new business opportunities, the early recognition of market discontinuities, and emerging dominant designs or more entrepreneurial approaches. Hence it can allow the parent company to identify and monitor new opportunities, and to externalize some R&D activities.

Investing in new ventures can also help the parent company to bring about internal cultural changes, by exposing management to entrepreneurship and new types of collaborations and synergies.

But developing CVC initiatives requires dedicated networking, negotiation, due diligence, deal making and business development skills which are often difficult to attract, develop and retain in large organizations. Indeed, while internal projects can be difficult to set up and manage, external ones can be even more challenging (see Chapter 4).

First, CVC initiatives might involve dedicated incentive schemes (such as stock options or success fees) that are difficult to implement within a corporation. As a consequence, such initiatives often fail because of a lack of entrepreneurial talent and difficulty in attracting and hiring skilled fund managers.[10] Developing strong networks and ties with venture capital communities can therefore be critical.

Second, the management of a portfolio of investments in new ventures requires a careful balance between financial objectives (investors' return) and strategic objectives (leveraging, option building and learning benefits). Indeed, most independent venture capitalist already struggle to achieve sustainable financial return, and adding a strategic dimension can create further complexity and ambiguity. In particular, it might be tricky to decide when an investment that is not financially profitable needs to be maintained for strategic reasons.

As a consequence, CVC initiatives often fail because parent firms are not able to deal with the financial risks and volatility involved in the management of such fast-paced environments, or because they fail to define and act according to clear strategic objectives. CVC initiatives also often fail because of a lack of sufficient autonomy and corporate commitment.

Third, the benefits and synergies outlined above suppose an effective integration of the new venture with the corporate culture and processes of the parent corporation, as well as effective collaborations with its business units. However, new ventures might not be ready to conform to such a culture, and business units might resist providing outsiders with access to their corporate jewels such as customer relationships, brand, R&D and manufacturing capabilities or cash. Ensuring strong complementarities in resources, adequate access to capabilities as well as positive relationships and communications is therefore critical.

As a consequence, CVC initiatives often fail to deliver synergies because of conflict of interests between the parent and the venture, caused in particular by the ventures' fears of expropriation by established firms[11] or difficulties in dealing with legal problems, such as the management of IP rights.

In other words, having a small firm working too closely with a large one can make it lose the benefits of smallness; while working not closely enough can make the parent lose the expected benefits of working together.

All these limitations mean that, while CVC initiatives might be promising – in particular when considering exploration strategies – few firms have succeeded in developing and implementing them successfully across business cycles. In particular, CVC initiatives are often effective only in industrial environments that combine weak intellectual property protection with dynamism and high technology potential, and among firms that combine high expertise in their own technologies and good entrepreneurship skills. In other words, CVC initiatives are a way to further improve the innovativeness of organizations, and not a substitute for it.

Key points to take away from Chapter 5

R&D activities are an important source of innovation opportunities for many firms. An effective R&D strategy is expected to define the technology platforms that will support the core competences of the organization and to manage resources across business cycles, combining bottom-up (project assessment) and top-down (strategic priorities) approaches. An effective R&D organization supposes to align the firm's strategy with the level of formalization, centralization and globalization of its R&D activities.

However, maintaining an effective R&D capability is not a sufficient condition for success, and R&D spending is in general not correlated with business performance. Firms must therefore leverage the sources of innovation opportunities that reach far beyond in-house R&D.

These sources of innovation opportunities include challenges to existing practices or plans, evolutions in market and corporate environments, and revolutions in business processes and technologies. They allow firms to leverage opportunities both inside and outside their organizations, scanning not only for raw ideas but also for more mature concepts that can be replicated successfully. Indeed "proudly-found-elsewhere" can often be more effective than "invented-here".

In particular, firms need to carefully balance "push" and "pull" approaches, in line with their strategy. "Push" approaches leveraging internal ideas and technology trajectories are often more risky but can unlock significant latent value, while "pull" approaches based, for example, on lead users, can be more effective but are often short-sighted.

In order to better harvest internally their corporate knowledge, innovative firms need to implement effectively and combine various types of search behaviors across their organization, in particular developing a network of gatekeepers.

They must also develop specific processes in order to manage and absorb both the explicit and tacit forms of knowledge created by their people, their structures and their markets. Moreover they should implement adequate knowledge protection mechanisms, identifying and dealing with the benefits and drawbacks of intellectual property rights and of alternative approaches.

Finally, innovative firms must design and implement externally specific intelligence and opportunity gathering processes. This includes better leveraging their market needs through value gap analysis, allowing firms to rebalance and expand their value propositions. It also includes better leveraging their industry dynamism by getting actively involved and tapping the innovativeness of new ventures through careful corporate venture capital initiatives.

Further reading

Amit, R., Brander, J. and Zott, C. (1998), "Why Do Venture Capital Firms Exist?", *Journal of Business Venturing*, Vol.13, pp. 441–466.

Bontis, N. (2001), "Assessing Knowledge Assets: A Review of the Models Used to Measure Intellectual Capital", *International Journal of Management Reviews*, Vol. 3, pp. 41–60.

Chesbrough, H. W. (2002), "Making Sense of Corporate Venture Capital", *Harvard Business Review*, March, pp. 90–99.

Cohen, W. M. and Levinthal, D. A. (1990), "Absorptive Capacity: A New Perspective on Learning and Innovation", *Administrative Science Quarterly*, Vol. 35, pp. 128–152.

Coyne, K. P., Clifford, P. G. and Dye, R. (2007), "Breakthrough Thinking from Inside the Box", *Harvard Business Review*, December, pp. 71–78.

Drucker, K. (1993), *Innovation and Entrepreneurship*, New York, Harper Business.

Grandstand, O. (2005), "Innovation and IPR", in J. Fagerberg, Mowery, D. and Nelson, R. R. (eds), *The Oxford Innovation Handbook,* Oxford University Press, pp. 278–285.

Grant, R. M. (1996), "Toward a Knowledge-Based View of the Firm", *Strategic Management Journal,* Vol. 17, pp. 109–122.

Greve, H. R. (2003), *Organizational Learning from Performance Feedback: A Behavioral Perspective on Innovation and Change*, Cambridge University Press.

MacMillan, I. C. and Block, Z. (1995), *Corporate Venturing*, Boston, MA, Harvard Business School Press.

Maula, M. V. J. (2007), "Corporate Venture Capital as a Strategic Tool for Corporations", in H. Landström (ed.), *Handbook of Research on Venture Capital*, Cheltenham, Edward Elgar, pp. 371–392.

Nambisan, S. and Sawhney, M. (2007), "A Buyer's Guide to the Innovation Bazaar", *Harvard Business Review*, June, pp. 109–118.

Saban, K., Lanasa, J. and Lackman, C. (2000), "Organizational Learning: A Critical Component to New Product Development", *Journal of Product and Brand Management*, Vol. 9, pp. 99–119.

Sandmeier, P. and Gassmann, O. (2006), "Extreme Innovation", *European Business Forum*, Vol. 26, pp. 44–49.

Schoenberg, R. (2003), "An Integrated Approach to Strategy Innovation", *European Business Journal*, Vol. 3, pp. 95–103.

Shaikh, J., Aafjes, M. and Bensaou, B. M. (2005), "Recharging Mobile Innovation", *INSEAD Quarterly,*, Vol. 11, pp. 30–35.

Smith, K. (2005), "Measuring Innovation", in J. Fagerberg, Mowery, D. and Nelson, R. R. *(*eds), *The Oxford Innovation Handbook,* Oxford University Press.

Tidd, J., Bessant, D. and Pavitt, K. (2001), "Learning through Corporate Ventures", in *Managing Innovation*, 2nd edn, Hoboken, NJ, John Wiley, pp. 279–312.

von Hippel, E. (1988), *The Sources of Innovation*, New York, Oxford University Press.

CHAPTER 6

Assessing innovations

Short case: raising funds for GeneticTools[1]

Dr D. was delighted when he learned that his home university had accepted his proposal to transfer the intellectual property rights related to the new molecule he had discovered. Laboratory tests had shown that this molecule could cure a special kind of liver disease affecting millions of people across the world. With exclusive access to such a technology in hand, Dr D. was convinced that investors would queue to finance his new venture. A systematic assessment of his innovation, however, revealed several important issues, which are summarized here.

First, having identified a large potential number of users was not enough. It was not clear whether the molecule identified could survive all the regulatory hurdles it faced, even if it could actually cure people. The readiness of patients and doctors to adopt this revolutionary treatment was also questionable, in view of the marketing power of existing competitors and the inertia of healthcare systems in many countries. Finally, many potential users could not afford to pay for expensive new medicines. The amount of revenue that could in reality be generated from the large potential number of users identified had therefore to be further explored.

Second, it is a long way from developing a molecule in a laboratory to producing and selling a drug worldwide. How it could be delivered, in which form and at what price was completely unknown. Furthermore, Dr D. had absolutely no experience at managing the complex manufacturing and

$\rightarrow$

sales process of a pharmaceutical business. It was therefore necessary to further validate whether a product based on the molecule could be put on the market, and how.

Third, a biotech start-up is never a one-man show. Dr D. had to demonstrate his ability and readiness to build a strong management and scientific team, beyond his former academic colleagues. Furthermore, it was clear that developing a business out of this molecule would include recruiting dozens of technicians and experts as well as establishing strong partnerships with international players. Dr D. was a recognized scientist but his abilities as a manager and deal-maker were still untested.

Finally, bringing a molecule to market involves significant investments and risks, which had to be better understood. How much was at stake and for how long investors had to be ready to commit funds was unknown. Finding investors ready to take on this type of risk and negotiate a deal with them would also be tricky, given the recent track record of similar ventures. In particular, Dr D. was convinced that he could negotiate significant royalties from future sales, but early investors were skeptical.

The systematic assessment of this innovation led to a better definition of the business model of the venture. It would focus on taking the molecule through the first phase of regulatory approval and then find a strategic partner. The assessment also led to a detailed human resource plan and to the enrolment of experienced board members. Finally, the financial valuation also highlighted the importance of negotiating attractive milestones payments in advance from the strategic partner rather than haggling about future royalties. As a result of this systematic assessment, Dr D. successfully negotiated a first round of financing worth several million euros.

However, a few years later the venture failed to raise any further finance, mainly because of personal conflicts between the management team and some key shareholders – something even a thorough assessment had not anticipated.

Having the capability of identifying innovation opportunities can be useless if there is no effective way to filter and prioritize those opportunities. This means that firms need to develop the capability of distinguishing good from bad opportunities, and to identify among the good opportunities the ones that will best support their strategic objectives.

In this chapter we shall discuss how this can be done at the individual opportunity level, through business planning. How the individual opportunities identified as being attractive can collectively form a balanced portfolio of projects will be addressed in Chapter 7.

This chapter focuses on radical innovations, where new resources must be mobilized and managed. In the case of incremental innovations, existing performance indicators such as cost, speed or quality can in most cases be used to assess whether an innovation opportunity should be pursued.

Assessing the value or the potential of an innovation opportunity supposes an understanding of (i) what business planning is about, what the key aspects are that need to be analyzed, in terms of both (ii) opportunity and (iii) resources, and finally, (iv) how to deal with the underlying uncertainties. These four aspects will be addressed below.

Introduction: the art of business planning

Faced with a radical innovation opportunity, whether it is a new concept identified internally or something already developed by a third party, a firm must assess whether and how much this opportunity should be pursued. The resulting assessment process should have several purposes.

The purposes of an assessment

The first purpose of the assessment process is, on the one hand, to minimize the risk of wasting resources, avoiding the pursuit of bad opportunities, and on the other hand to prioritize the utilization of those resources, as firms want to allocate their limited means to the most promising prospects. The assessment process must therefore be based on a systematic analysis of all the relevant aspects of an opportunity.

In particular, business plans often share similar strengths and weaknesses. In the same way that inventors tend to focus on newness and forget the importance of change (see Chapter 1), poor business plans often focus on product and technology and do not sufficiently address the issues of marketing and distribution. In contrast, good business plans understand the importance of mobilizing resources and cover convincingly the issues of competitive products and services (not only technology), target market segments, distribution channels, adequate funding, competent teams and business partnerships.

We must stress that this purpose also means that an assessment ultimately concluding that an opportunity is not worth pursuing is certainly not necessarily a wasted effort. In other words, this is about validating an opportunity. But the assessment should also have other purposes, detailed below.

The second purpose of the assessment process should be to refine and improve the opportunity. As information is gathered and hypotheses are developed regarding various aspects of an opportunity such as the offer, the market or the competition, the initial assumptions about what the innovation opportunity was about should be revisited and challenged, adjusted and improved as necessary. In other words, this is about fine-tuning, about learning and building know-how. To paraphrase Dwight D. Eisenhower, this is about realizing that "plans [might be] useless, but planning is indispensable".

Hence the assessment process itself matters as much as the end product, as it is an opportunity for the organization and/or the team involved collectively to develop new knowledge and decrease uncertainty. This also means that assessment should be managed as an iterative process, where what is learned at each step can challenge what was decided during previous steps.

The third purpose of the assessment process should be to provide legitimacy to the opportunity and facilitate the enrollment of key stakeholders (see Chapter 4 regarding the importance of enrollment). Even if a manager or an entrepreneur is convinced about the value of an opportunity, telling colleagues "just trust me" or to do it because "I said so" can be poor ways of convincing and engaging others effectively and mobilizing resources.

In particular, the assessment process should be based on an approach that is perceived as legitimate and credible by the relevant stakeholders (colleagues, managers, investors, suppliers or partners) and communicated to them in a suitable way.

In other words, this is about convincing others, about making up people's minds about an opportunity.

Business planning

The assessment process described above is usually called business planning. Let us stress that when considering innovation opportunities this concept is misleading, as it is not really about planning. Indeed, as discussed above, the assessment process is about analyzing and validating, about refining and learning, and finally about convincing and legitimating an opportunity, in order to decide whether or not to mobilize resources to pursue it.

In particular, the main objective of the business plan of an innovation opportunity is not to predict or guess what will happen in reality. This is obviously impossible, not only because of the high uncertainty involved but also because business plans are endogenous. In other words, what will happen is influenced by what the firm will decide and do, which itself is influenced by how the assessment process is conducted.

As a consequence, business planning is not about predicting. It is rather about trying to identify whether credible success scenarios can be built, and to gather arguments and evidence in order to convince others about the likelihood of such scenarios. It is about trying to outline in a professional way a vision of a possible and attractive future, not about forecasting events.

This explains why many entrepreneurs claim that they never look back over their business plan, or that what was in the plan was far from what actually happened. However, this does not mean that completing the plan in the first place was not useful.

Another important aspect is that there should be no such thing as a universal business plan template. What the issues addressed in the plan are will depend on the nature of the opportunity (such as industry, technology, size, maturity or risks). How those issues are addressed will depend on whom the team or organization involved try to engage or convince.

Indeed, a business plan is also in many cases a communication tool. As such, its content and format will depend on who it is aimed at (managers, investors, bankers, partners or team members), and what the objective regarding this audience is (to engage, to convince, to learn and so on). In many situations, "plans are developed and submitted to resource controllers to obtain the resources needed to launch innovation development. In most cases, the plans served more as 'sales vehicles' than as realistic scenarios of innovation development."[2] As an example, a business plan written in order to convince a management committee is going to be very different from the same business plan written to secure a loan from a bank.

The content of each business plan will obviously be linked to the specific characteristics of the innovation opportunity considered, in terms of market, industry, technology or maturity. However, there are issues that should be addressed in all business plans, questions that always need to be dealt with. This is particularly the case at the early stage of development of an innovation opportunity, when the focus is often too much on the technology alone.

We stress again that business planning is not about forecasting but about gathering facts and arguments in order to make an educated decision. Moreover, this is an iterative process, where the value lies in the process itself as much as in the end product. While the output of that process will vary depending on the context of the opportunity and the objectives of the business plan, there are core questions that always need to be addressed.

Those core questions are about both the innovation opportunity itself, why it exists and what could be done about it, and the resources needed to capture it – in other words, who to involve and how much financing is needed. Those aspects (see Figure 6.1) will be addressed below.

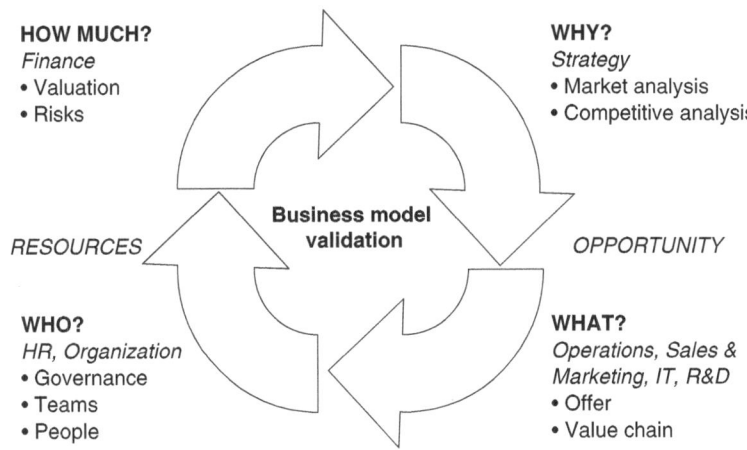

HOW MUCH?
Finance
• Valuation
• Risks

WHY?
Strategy
• Market analysis
• Competitive analysis

RESOURCES

Business model validation

OPPORTUNITY

WHO?
HR, Organization
• Governance
• Teams
• People

WHAT?
Operations, Sales & Marketing, IT, R&D
• Offer
• Value chain

Figure 6.1   The business planning cycle

Opportunity: why it is there and what it is about

The first core question to address regarding the assessment of an innovation is whether that innovation indeed represents a credible and sizable opportunity, both from a strategic perspective and a business model point of view.

The strategic perspective concerns, on the one hand, whether there is in fact an actual opportunity (why?) and on the other, whether the firm could be well positioned to capture it (why me?). From a strategic point of view, a firm considering an innovation opportunity must indeed first analyze, learn and legitimate whether there is a significant value creation potential and then whether the firm enjoys some competitive advantages with regard to this.

In other words, it is about justifying the attractiveness of the opportunity (why is it a good opportunity?) and its relevance for the firm (why is it the right opportunity for me?).

Many innovative technologies and niche markets represent opportunities that are not attractive, or not attractive enough from a business point of view. Similarly, an innovation opportunity that is attractive for one firm will not necessarily be attractive for another, because the latter faces a different environment, has access to different resources and/or pursues different objectives.

As an example, an opportunity to develop a new online or offline payment system will be considered differently by a bank, a credit card company, a telecom company or an online retailer, among others, because they each face a different competitive environment and can leverage different assets – which might be their infrastructure, expertise, brand or customer base.

Having addressed the strategic questions regarding the market potential of the opportunity (why?) and its competitive fit (why me?), the next question to address is the business model, considering the value proposition (what will be sold and to whom?) and the value chain (how will it be delivered?). Identifying an innovation opportunity that makes sense from a strategic point of view (there is a sizable opportunity and the firm is well positioned with regard to that opportunity) is not enough. What is needed is an actual way to develop a business activity – in other words, a business model.

Ways of dealing with the two strategic issues and the business model will be addressed below

Strategy: why there is an opportunity

Assessing why there could be an opportunity means first understanding what the needs are that the innovation could meet, what the difficulty (the "pain") is that it could address or the job it could fulfill. This can be done, for example, through interviews, focus groups or anthropologic research, observing how people actually behave in specific situations. This can relate either to internal users or colleagues when considering a process innovation, or to external users or customers for a product or service innovation.

In both cases, the needs identified typically include basic physiological or safety needs (such as health, nutrition, wealth or energy) but also more social needs related to belonging, esteem or self-realization. In particular, most people will not adopt an innovation just because it is new, but some will value newness itself because of the social image it reflects.

The second aspect of the "why?" question is defining who the groups of users (segments) most likely to adopt the innovation are, and how numerous they are, based on their total population, social characteristics (age, wealth, geography and so on), and their aspirations or behaviors. While many techniques exist to assess the size of a potential market, we shall outline here two generic approaches, including both top-down and bottom-up points of view.

A top-down assessment begins with an evaluation of the total number of all users who might be interested, now and in the near future. This total number, or potential market, is then broken down, considering in particular:

- The share of users who could (in theory) adopt the innovation if they were aware of it, because they could afford it, they could have access to it and they would be allowed to do so. For example, many new drugs are only available in rich or subsidized markets, or only through prescriptions.
- The user segments targeted by the innovation, which are the users that could actually want to adopt the innovation because they are aware of it and because it is available to them. For example, many new technologies are initially deployed and/or marketed only in specific regions.
- The individual users who will decide to adopt the innovation and those that will choose the offer of the firm among the competing alternatives. For example, many people have not yet adopted energy-saving products or behaviors, though they could, and would have to choose among all the available solutions.

The typical result of such top-down approach estimates the potential as the average monetary value of the need, multiplied by the number of adopters. The number of adopters itself is estimated as a fraction of the potential market, corresponding to the users for which the innovation is available who are targeted by the firm, who actually adopt the innovation and who choose the firm's offer.

It must be stressed that such an approach can be dangerous to use, as it accounts only for theoretical users who might not actually exist. Furthermore, a small percentage applied to a sufficiently large number can always appear to justify a sizable potential. As an example, many business plans completed during the dot.com bubble at the end of the 1990s were based on the huge growth in internet usage but referred to customers who never materialized.

As a consequence, such a top-down approach should be combined with a bottom-up assessment. A bottom-up assessment starts with the identification of actual users who have adopted the innovation and/or plan, are committed to adopt it, or are likely to do so.

The result of such an approach is a list of identified adopters or likely adopters, gathered from among lead users, test group members and/or prospects. Such an approach goes beyond desk research and confronts the innovation opportunity with real-life potential adopters. It also reveals not only who could want to adopt the innovation but also who actually makes the decisions, who mobilizes the resources needed (who pays?) and/or who can influence the decision (such as hierarchy, advisors or relatives) to adopt the innovation or not.

The bottom-up approach allows the strengthening and validation (or not) of a top-down assessment of the number of virtual users, by identifying actual potential users. In particular, a bottom-up approach allows a better knowledge of who the potential users could be and is aware from an empathic point of view what is the perceived value-added for them and what are the motivations that could convince them to change.

As an illustration, many research-based start-ups trying to serve business customers, and many existing firms trying to venture into new sectors, fail because they do not have a good understanding of their target customers' behaviors and expectations. Similarly, many entrepreneurs targeting consumers overestimate the appetite of other people for their products or services, often by projecting their own preferences on to them.

As discussed in Chapter 5, it is often very difficult to predict how people will react to an innovation. The question is, therefore, not to try to guess whether each potential user will actually adopt an innovation, but rather to gather arguments and evidence about whether they could credibly do so, in order to better decide, learn and/or convince.

Indeed, while most predictions about the rate of adoption of an innovation are eventually proved to be quite wrong, firms who innovate successfully are often those who got those predictions less wrong than others and/or learned from their mistakes and adapted faster.

Strategy: why it is for me

Having identified a sizable need that could be met, and the resulting value creation potential, is a necessary but not sufficient condition for a firm to deem an innovation opportunity to be attractive. On top of analyzing, learning and legitimating "why" there is an opportunity, firms must also decide whether that opportunity is attractive *to them*. They must consider whether the opportunity fits with their strategy and whether they enjoy sources of competitive advantages that they can leverage in order to capture that opportunity better than others.

In other words, firms need to find out not only whether there is a big cake out there, but also whether they are well positioned to eat it. There have been many cases where a firm that initially uncovered an opportunity (as an inventor or a first mover) failed to capture its value creation potential and was overtaken by others. Conversely, a patent (see Chapter 5) is a way to motivate innovators by putting them in a good position to exploit the opportunity they identified.

A sizable innovation opportunity will be attractive for a firm if it fits its strategy and if the firm can differentiate it from its potential competitors. We shall discuss both these aspects below.

As discussed in Chapter 3, the fit with the strategy is driven by the resources (feasibility, strengths and weaknesses), the environment (suitability, threats and opportunities) and the purpose (vision) of the firm.

In terms of resources, the "why me?" question is used to analyze, learn and legitimate whether, in particular:

- The firm has or could build the technical, organizational and leadership capabilities required to capture the innovation. Many utilities have, for example, struggled when trying to develop customer-oriented value-added services.
- The economics of the current business model and the available funding sources can support the required investments. Many oil companies are, for example, well positioned to capture opportunities related to alternative energies.
- There are underutilized assets that the firm could leverage. Retailers, for example, are using their available space and customer traffic to develop new businesses.

In terms of environment, addressing the "why me?" question is used to analyze, learn and legitimate whether, in particular:

- There are capabilities available in the firm's ecosystem (see Chapter 4) that it could leverage. Locating activities within a technology cluster can, for example favor the capture of innovation opportunities.
- The industry structure favors the emergence of the innovation considered, both in terms of potential profit pools and competitive pressures. Many new media businesses are, for example, struggling to be profitable, given the sector's competitive intensity and the pressure from suppliers and alternative providers.
- The industry maturity and socio-economic environment favors the emergence of the innovation considered, in terms of the rate of technological change, cultural evolutions, regulations and convergence or divergence. For example, many new services developed by cell phone handset manufacturers integrate the technology evolutions of chip manufacturers as well as the development of new regulatory standards (such as the 3G or 4G mobile telephony standards).

In terms of purpose, addressing the "why me?" question is used to analyze, learn and legitimate whether, in particular:

- The impact, both short- and long-term, of the innovation opportunity on the performance of the firm and its alignment with the aspirations of its main stakeholders. For example, many industries pursue innovation opportunities reducing their carbon footprint because of both financial and social pressure.

- The main stakeholders are ready to face the risks and uncertainties linked with the innovation, both in terms of financial results and reputation. For example, new complex products or venture capital investments are much less popular among shareholders in the aftermath of financial crises.

But on top of the fit with its strategy, a firm must also consider whether it enjoys strong sources of competitive differentiation regarding the innovation opportunity. In other words, whether it will able to capture this opportunities in ways others will have trouble emulating.

This is especially critical for start-ups or for firms entering new sectors of activities, as they often fail to anticipate or deal with the reactions of incumbents or of other firms in their value chain that could leapfrog them. If a firm has identified a great opportunity, it is indeed foolish to assume that other players will watch it develop and not react.

While there are many ways a firm can try to develop a differentiated offer, we can highlight three sources of competitive differentiation that are particularly relevant when considering innovation opportunities. These are:

- The firm can enjoy rents or economic profit from its market power in the industry considered. For example, large retailers or large firms in the information technology sector can introduce innovations in ways that are inaccessible to other players, in terms of both reach and cost structure.
- The firm can enjoy rents from unique assets relevant to the innovation being considered. While the obvious example is a patent, other unique assets can relate to infrastructure (for example, for a telecommunications operator) or brand.
- The firm might not necessarily enjoy the benefits of scale or unique assets but can identify and capture opportunities much faster than others. By the time their main competitors have replicated the innovation, such firms have thoroughly

"milked" the market and are already moving on. There are many examples of new models or products launched in the auto industry or in the fast-moving consumer goods (FMCG) sector that eventually were copied but yet allowed the innovating firm to create temporary competitive advantages.

Hence a firm considering an innovation opportunity must analyze, learn and legitimate whether it could leverage at least one of these sources of competitive differentiation. If not, it needs to consider whether it has any chance of building a significant market share and enjoy decent margins.

Having discussed the two strategic aspects of the assessment of an opportunity (the innovation potential and fit), in the next section we shall address the issue of designing a credible business model.

Business model: product/market positioning and value chain

Designing a business model means defining, on the one hand, a realistic value proposition, and on the other a feasible value chain. Indeed, having identified an innovation opportunity, a firm must define how it could capture it in a way that makes economic sense. This means first defining a value proposition, by defining what is an actual offer that could be sold and what are the attributes it will need to have to convince potential users to adopt it.

As an illustration, there were many firms other than Google that had identified the opportunity to develop innovative search engines (in other words, there was a sizable opportunity), but Google is the one that developed a business model (simply offering a search service and selling ads next to search results) that succeeded in leveraging that opportunity in a profitable way. Conversely, many people would like the environment to improve (there is perhaps a need) but that does not mean that they are ready to pay for this (so there is perhaps no business).

A simple way to define what designing an offer entails is to compare it to conceiving what the first invoice sent to early adopters of the innovation could look like. An invoice must specify in an unambiguous way what the price is and what exactly is being paid for (for example, a certain quantity of the product or access to a specific service). Invoicing also defines where the offer will be delivered (the place) and in which format (the packaging). Finally, an invoice is sent only because a potential user was made aware of the offer (the promotion).

Indeed, users or customers do not adopt innovations, concepts or technologies. What they do is to purchase actual offers, defined by their product, price, place, promotion and packaging (the 4P+1 of marketing). Obviously, in the early stages one will not know exactly how such a value proposition will be implemented. However, it is critical to analyze, learn and legitimate whether there are ways in which it *could* be implemented.

The design of this value proposition must take into account in particular that adoption (or purchasing) is a complex process, and that it goes well beyond the simple acquisition of an object or the use of a service. What constitutes an effective unique selling proposition (USP) is not driven only by what the users will actually pay for and the minimal features provided. The adoption process will also be strongly influenced by what the adopters actually believe they will get (the expected attributes), by additional features beyond what is expected, and by future features beyond what is currently offered (the potential attributes).

As an example, what is true for a new automobile or a new cell phone (that most consumers do not buy only mobility or communications when they purchase such items) must also be taken into account when developing a new drug, a new manufacturing process, new software or a new source of energy. In particular, a single innovation (such as mobile telephony or social networking) can lead to a wide variety of value propositions, some more successful than others.

Having assessed how an actual offer could be designed, the second aspect of the business model that needs to be considered is the value chain; in other words, how that offer could actually be delivered.

The technical aspects of such a value chain, how formally it is managed and how it can be implemented will obviously vary from one opportunity to another. However, there are underlying elements that always need to be present (see Chapter 3). Whatever the innovation opportunity relates to, delivering the offer to capture it will always require the implementation of design, operations, client management and support activities.

First, the design activities relate to how the firm will ensure that the value proposition identified is conceived, updated and upgraded effectively, both initially and as the project matures. It relates to the "sourcing" capability discussed throughout Chapter 5, and in particular to the way that the firm will take care of the key elements of a value proposition (the 4P+1 mentioned above).

The team or organization involved in the innovation opportunity will have to decide, for example, whether the firm will develop specific R&D or marketing activities, whether the product will be customized or standardized, relatively stable or constantly improved, and how that will be done and by whom.

Developing a sustainable business model involves analysis, learning and legitimating ways of addressing these design issues. Failing to do so can lead to a value proposition that quickly become obsolescent or with only a very limited adoption.

The next key elements of a value chain are the operations activities. These relate to how the firm will ensure that the product or service is delivered in an effective way to each user, from the initial adopter to the mainstream user. Obviously, this includes a consideration of how the product or service will actually be manufactured, produced and assembled, and which supply management and operational processes will be put in place.

But addressing operations activities also means dealing with issues such as capacity scale-up and resource bottlenecks: how to deliver effectively one, then a hundred and perhaps one day 10,000 times the original product or service. It also means developing hypotheses regarding the location of the activities, based on where the end-users, the suppliers or the expertise needed are located. It means finally analyzing, learning and legitimating optimal workflow management (batch or flow, just-in-time or flexible) as well as quality management and certification processes, in line with relevant regulations.

The third key element of the value chain relates to client management. It concerns all the interactions that must take place before and after potential or actual users adopt the product or service. It relates to how the firm will ensure that marketing and communication activities are developed, that the sales process itself can be dealt with (for example, through intermediaries, own points of sale, franchise or direct sales), and that adequate customer or user support can be ensured.

We must stress the importance of these client management activities, as insufficient resources and poor marketing and distribution planning are a very common feature of failed business plans, in particular in the case of high-tech start-ups and spin-offs. Similarly, innovative processes designed in one part of an organization often fail to be adopted because of insufficient follow-up, lack of customization or poor technical support.

The fourth key element of the value chain relates to support activities. It concerns the back-office or overhead activities required by the design, operations and client activities, such as human resource administration, accounting, public relations or legal services. While effective support services are definitely not a sufficient condition for success, they are often a necessary one and can divert much scarce management attention if not dealt with properly.

In summary, assessing an innovation opportunity requires the assessment from a strategic point of view of whether it could be attractive for the firm but also to analyze, learn and legitimize whether a successful value proposition and an effective way to deliver it could be put in place. This means developing credible hypotheses regarding what could be a potential offer in terms of product, price, promotion, place and packaging, and considering what could be a potential value chain in terms of design, operations, client management and support activities.

As an example, the failure of many promising start-ups can be linked to their inability to design and deliver effectively a product or service that many users will actually pay for. Conversely, the success of innovation icons such as Apple or 3M can be linked as much to their product design and operational effectiveness than to their inventiveness or strategic vision.

Having assessed the opportunity from both a strategic and business model point of view, the second facet of the assessment of an innovation opportunity concerns the human and financial resources needed to capture this opportunity. These will be discussed in the next section.

Resources: who and how much?

The two scarce resources that can hinder the capture of even the most attractive innovation opportunities are people and money. Assessing whether an innovation opportunity should be pursued therefore requires analysis, learning and legitimating whether the right organizational capabilities can be mobilized, and whether the expected benefits of the opportunity are greater than the costs it involves. Both aspects will be discussed below.

People: who will do the job

As discussed in Chapter 1, innovations are the product of organizations, not lonely inventors. Assessing an innovation opportunity therefore means considering whether the right people and organizations can be mobilized to capture this opportunity. This means, in particular, mobilizing the right management team and the right partners, and designing the right governance structure.

Indeed, the lack of managerial capabilities has long been known to be a major brake on the development of new businesses, and many project failures can be traced back to unbalanced or weak project teams (see Chapter 4).

Again, while the initial assessment of an innovation opportunity does not require the identification by name of all the people that will be involved, it is critical to consider whether the right managerial skills and organizational structure could be mobilized. This is particularly the case for scarce skills or critical capabilities.

The first task is therefore to consider whether an adequate entrepreneurial project team could be put in place and developed, leveraging available people or hiring new ones. As discussed in

Chapter 4, such a team should balance the various executive, financial, technological and marketing experience needed, combined with leadership and networking skills as well as industry and market expertise. The way these people will be motivated and rewarded should also be considered; in particular, the individuals whose involvement is critical to the success of the project as well as those which are expected to contribute only part-time (and therefore need to arbitrage other commitments).

The second task is to consider how to organize, initially and as the opportunity matures, each of the four key activities of the value chain described in the previous section. In other words, to assess who could manage the design, delivery, client management and support activities required to capture the innovation opportunity. This means knowing for each set of activities what are the key success factors (such as quality, cost or reliability), and whether they could best be developed in-house or through partnerships.

In particular, this means considering which steps of the value chain relate to critical competencies that need to be developed in-house and which are best outsourced, in order to decrease cost, risk and delays or to access critical expertise or markets (see Chapter 4). Many new businesses fail or struggle because they did not engage the right partners in the early stages, either through ignorance or because they wanted to do everything themselves.

Having considered who could do what, the third task is to assess which effective governance structure could be put in place to pilot the capture of the innovation opportunity. Various statutory or legal organizational structures may be considered, taking into account the regulatory and fiscal environment as well as the stakeholders involved. The governance structure should address who should provide strategic guidance and legitimacy (such as a board or a steering committee), who should provide specialized skills (subcontractors, partners or networks), and how those two entities will interact and share information with the management team in charge of operations and performance. Many entrepreneurs or project leaders actually fail to recognize the strategic importance of a board of directors or of a steering committee, in terms of coaching, credibility, sponsoring, networking and support.

In summary, assessing the resource from a people point of view means considering who (individuals or organizations) could best perform each of the core activities required to capture the innovation opportunity, both at the initial stage (launch) of the project and as it matures.

As an illustration, an innovation opportunity involving a regulatory expert, a patent lawyer or a software guru might be delayed or discarded just because such profiles are unavailable or too expensive to recruit. Similarly, the lack of managers with strong business development skills can hinder the ability of a firm to capture multiple innovation opportunities.

As well as people and partners, capturing attractive innovation opportunities also require mobilizing financial resources, either to fund investments or to justify the opportunity cost of using some available resources (people, factory or distribution channels, say). These will be discussed below.

Finance: how much will it cost, or earn?

Having identified a relevant and sizable opportunity, designed a business model to capture it and considered the people and organizations that needed to be mobilized, the final aspect of the assessment of an innovation opportunity is to analyze, learn and legitimate whether the effort will in fact be worthwhile.

In other words, it means assessing for each period of time considered whether the potential revenue and/or cost savings expected from the innovation opportunity are likely to be greater than the fixed and variable costs of the resources mobilized, given the strategy, the business model and the organization (see Figure 6.2).

While the underlying calculations are in most cases quite simple, the accuracy of the financial assessment will obviously depend on the quality of the assumptions made regarding the various sources of cost and revenues. In other words, this process can be summed up by the phrase "garbage in, garbage out".

A good assessment of the strategic potential (why?) and the differentiation (why me?) must guide the assumptions made

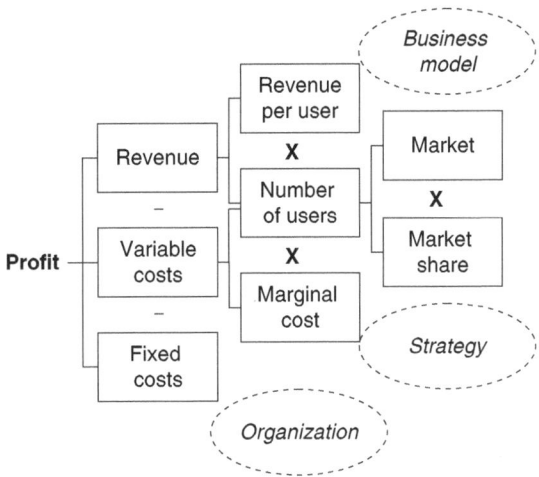

Figure 6.2 Fundamentals

regarding the potential market and market share, and hence the number of adopters (demand or volume).

A good understanding of the value proposition (what?) must guide the assumptions made regarding the average revenue per user (average price or average cost saving) and variable costs (including raw materials, energy, and sales and distribution).

Finally, a good understanding of the value chain (how?) and of the human resources mobilized (who?) must support the assumptions made regarding the fixed costs (including investments, overheads or marketing).

Once the expected benefits have been assessed over the lifetime of the innovation opportunity, it is possible to derive some key basic metrics regarding this opportunity, in particular an estimate of the market potential, of the pay-back period and of the upfront investment (see Figure 6.3).

The *market potential* provides a quantification of the size of the business opportunity considered, which can be measured in yearly cash flow or alternatively yearly gross margin, earnings or sales. This metric allows analysis, learning and legitimation of whether the opportunity is good enough. It allows in particular an examination of the size of the opportunity with

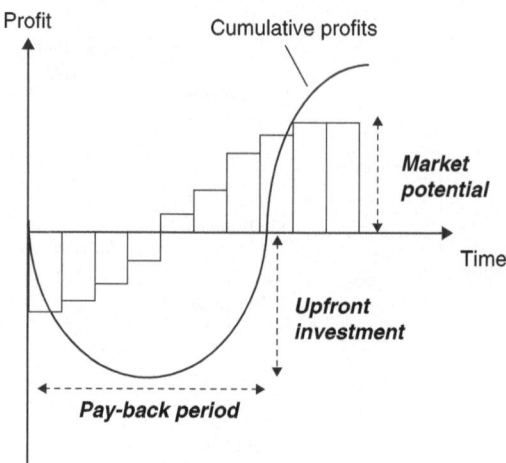

Figure 6.3   Key valuation metrics

the expectations of the stakeholders involved (management or investors, for example) and with the opportunity costs of exploring and/or pursuing the project, either in terms of management attention, mobilized assets or due diligence costs. As an example, developing a new type of glass design especially for aquariums is probably not worth the effort for a large glass manufacturer, while developing new glass aimed at protecting solar panels could be. Similarly, developing a new polymer to offer new plastic wine bottle stoppers is probably not worth the trouble for a large plastics manufacturer, but developing new polymers for fuel cells could be.

Second, the *upfront investment* is a measure of the total funding needs of the project, of all the financing that must be mobilized in advance to capture the opportunity. This amount must in particular be confronted with the ability and readiness of the stakeholders involved to provide or raise cash. As an illustration, developing a new automobile platform, a new generation of microchips, a new blockbuster drug or a new aircraft design can be very attractive but involve upfront investments of hundreds of millions of euros, which very few firms can afford. Similarly, some innovation opportunities can only be targeted if the firm is able to mobilize external sources of financing, such as public funds or supplier financing.

Finally, the *pay-back period* provides a quantification of the number of months or years expected to elapse before the opportunity might pay back its initial investment. This metric indicates whether the timing of the opportunity does match the objectives and relative patience of the stakeholders involved.

Setting a relatively short limit to the maximum acceptable pay-back period (for example, three years) can reflect mistrust in the reliability of any prediction beyond this time horizon, or the expected rapid obsolescence of the innovation opportunity. This time limit can be related to the extinction of a patent protection or to the technology lifecycle. As an illustration from micro-electronics, new generation of chips are expected to appear every eighteen months.

As an example, venture capitalists investing in start-ups typically expect to exit their investment after five to seven years. Firms struggling with solvency or under heavy pressure to deliver predictable quarterly financial results will often focus on opportunities with a very short time horizon. Conversely, firms with stable ownership and a long-time horizon can often accept longer pay-back periods. Indeed, acceptable pay-back periods can go from weeks in the computing, fashion and entertainment sectors to decades in the aeronautics or mining sectors.

Hence those three basic metrics can provide some valuable information regarding the attractiveness (or not) of the innovation opportunity considered. It is also possible to mobilize more advanced metrics (such as the return on investment – ROI, or the internal rate of return – IRR), bearing in mind, however, that it is pointless to use sophisticated valuation techniques if the reliability of the underlying assumptions is limited.

Probably the most popular of the more sophisticated valuation techniques is the discounted cash flow (DCF) or net present value (NPV) approach, which allows the conversion of the net cash flows identified (see Figure 6.3 above) into a single, risk-adjusted cash value. This technique offers many variations[3] but its basic principles and limitations will be discussed below.

The first principle of the DCF or NPV approach is that a risky euro tomorrow is less valuable than a certain euro today. Hence future cash flows are discounted each year. The discount rate reflects the opportunity cost of the capital mobilized, which increases with the estimated riskiness of the innovation opportunity. Indeed, riskier projects are expected to provide higher returns. This means that such an approach is risk-adjusted, while other metrics such as ROI or IRR are not.

Typical discount rates used for corporate projects range from 10 percent to 15 percent, while investors in high-tech start-ups can use rates of up to 25 percent to 30 percent, as a result of the inherently risky nature of such ventures.

The second principle of the DCF or NPV approach is to take into account all the future net cash flows linked to the innovation opportunity. By contrast, metrics such as the pay-back period or upfront investments consider only the initial cash flow.

The DCF or NPV approach requires on the one hand the discounting and summing-up of all the future net cash flows for which reasonable assumptions can be made, and on the other hand to estimate and discount the final value of the remaining cash flows (the "final" value). The value of the innovation opportunity is then equal to the sum of the discounted cash flows considered plus the final value.

Typically, cash flows are estimated over the first five or first ten years, but sometimes more in the case of innovations with very lengthy lifecycles, such as aircraft engines.

The final value can then be estimated either as zero (in the case of an innovation facing complete obsolescence), negative (in the case of an innovation involving rehabilitation or recycling costs, as, for example, in the energy sector) or as a proxy of future cash flow based on resale value, balance-sheet metrics or "perpetual" value.

There are three main sources of attractiveness of the DCF/NPV approach:

- It associates a cash value with an opportunity, rather than a time period or a relative rate.
- It allows the consideration of projects with different risk profiles; riskier projects being discounted more heavily.
- It does not involve setting an explicit arbitrary threshold such as a minimum rate of return or a maximum pay-back time.

However, the DCF/NPV approach also has strong limitations when it is used to assess innovation opportunities, which are often ignored or underestimated. These limitations are related to the two principles outlined above, which are the risk-adjusted value of cash-flows and the infinite time horizon.

First, the resulting value estimated for an opportunity is highly sensitive to the value of the discount rate used, which itself is somewhat arbitrary. Even when using common techniques to quantify the discount rate (such as the weighted average cost of capital), one has to make assumptions about the riskiness (the "beta") of the opportunity, which is often very difficult when considering innovation opportunities. The fact that the result can be very sensitive to these assumptions and that the assumptions are actually hidden in the calculation of the discount rate (and therefore often not discussed at management level) is an important limitation to using this approach to assess innovation opportunities. As a consequence, some investors and firms impose a fixed discount rate in advance to use for all valuations, to neutralize these considerations.

Second, the estimated value of a project is by definition the sum of the discounted cash flow considered and of a residual final value. While the first are the focus of the assessment and are relatively well known, the latter is relatively far in the future and therefore often much more uncertain. However, in many cases, the final value is far from being marginal and can represent an important share or even most of the total estimated value of the opportunity.

In other words, in some cases the estimated value is the sum of a relatively well known but small number (the discounted cash flows) and of another very uncertain but large number (the final value). As an illustration, this is like saying that the value of a firm is €4,123,236 when it is in fact €2,235,236 plus more or less €2 million.

As a consequence, the reliability of the DCF/NPV assessment of an innovation opportunity will depend not only on the quality of the assumptions made regarding the strategic opportunity, the business model and the resources mobilized (see above). It will also depend strongly on the quality of two further assumptions regarding the opportunity, which need to be analyzed, learned and legitimized. These two further assumptions are, on the one hand, the expected riskiness of the opportunity, which will influence the discount rate, and on the other its long-term profitability, which will influence the final value.

The main implications of those limitations are twofold. First, managers should not treat DCF/NPV approaches as straightforward black boxes with mathematical precision. They should be aware of the assumptions they are making, in particular regarding the discount rate and the final value calculations. Second, the DCF/NPV approaches are better suited to comparing and prioritizing multiple innovation opportunities, when one can take similar discount rates and final value assumptions. When valuing an individual innovation opportunity, such methods should be combined with other approaches, based, for example, on market multiples, asset values or comparable transactions.

In summary, given the strategic assessment of an innovation opportunity, a business model and the organizational implications, it is possible to analyze, learn and legitimate quantified metrics related to the cash flows linked to the opportunity. Used with care, these metrics can guide the management of the innovation opportunity by providing indications of its potential financial value. An estimate is perhaps often wrong, but it is better than a random number or no number at all to support a decision-making process.

Again, the process of valuing an opportunity is often more meaningful than its outcome. It can, for example, reveal that some strategic options actually destroy value, or that some elements of the business model are much more profitable than others (see the short case at the beginning of this chapter).

Furthermore, we stress that the decision for an existing firm of whether to exploit an opportunity will not depend only on the financial assessment of that opportunity. How an opportunity threatens or contributes to the existing activities of the business will also influence the decision, whether in terms of opportunity cost, cannibalization risks, strategic value or exposure to uncertainties.

Similarly, a financial investor such as a venture capitalist will not take into account only the potential value of a given deal, but also consider the intrinsic quality of the entrepreneurial team, the overall attractiveness of the market under consideration, and the fit between the opportunity and the characteristics and objectives of its fund (such as timing, scope, geographics, deal structure, or synergies with investor expertise and with other deals). Again, an innovation opportunity that is considered to be attractive for one firm or one investor might not be for another.

The valuation process can also contribute to highlighting whether the attractiveness of an innovation opportunity is particularly sensitive to the achievement of some key performance thresholds (such as a minimum market share or a maximum unit cost) or deadlines (such as obtaining a patent or concluding a partnership). The process can therefore highlight which performance indicators or which deliverables need particular management attention.

We shall discuss in the next section how the underlying uncertainty linked to these factors can be taken into account in the valuation process.

Integrating uncertainties

The newness of the nature of innovation opportunities generates intrinsic uncertainties and risks which should be taken into account when analyzing, learning and legitimizing whether they are worth pursuing.

Indeed, innovation opportunities often miss significant track records and business history. In most cases they lack the built-in confidence and expertise that could accurately support the assumptions made regarding those opportunities. Precautionary approaches and risk-aversion can therefore lead to adverse selection, valuable opportunities being rejected because of the uncertainty surrounding them. Conversely, unwarranted optimism and technocracy can lead to huge wasted investments.

Since Western Union (the leading telegraph company at that time) considering in 1876 that "the device [the telephone] is inherently of no value to us",[4] the history of technology teaches us to be very humble regarding our ability to forecast the success or failure of an innovation. The wide variations observed in venture capital investments and high-tech stock market valuations in recent decades provide other reasons to remain cautious.

Furthermore, exploring an innovation opportunity often involves sunk or unrecoverable investments in new capabilities. The knowledge and expertise developed in a new field and the specific assets funded to explore it can have limited value if the opportunity fails to materialize. Hence not only the uncertainty is higher than for traditional investments, but so are the risks taken. For example, a chemical company can make heavy investments in new, sophisticated laboratory equipment that will not be worth much to third parties. Similarly, the knowledge and expertise developed by automobile manufacturers in the field of hydrogen engines could become essentially worthless if this technology fails to reach any market acceptance.

As a consequence, it is critical to understand well what the sources of uncertainties involved in an innovation opportunity are, and what their impact on its assessment could be. These two issues are discussed below.

Sources of uncertainty: know that you do not know

The sources of uncertainties relate to competition, viability, management and investments risks, which can affect all the aspects of the business plan. First, competitive risks will affect the value of the opportunity ("why?") and the ability of the firm to capture it ("why me?"). They relate in particular to the benefits potential users will actually perceive, their readiness to pay and their resistance to change (see Chapter 1). These risks will also be linked to the perceived differentiation *vis-à-vis* alternative offers and to the way competitors and potential new entrants might react.

As an example, the market potential of satellite phones was widely overestimated, in particular because of an underestimation of the quick diffusion of land-based cell-phone infrastructures.

Conversely, the market potential for digital cameras and digital music players was widely underestimated by many incumbents, in this case the result of a failed understanding of the perceived benefits they could generate, such as mobility and customization.

Second, viability risks will affect the feasibility of the investments required by the business model. They relate in particular to regulatory constraints (such as product/market regulations and technology standards) and to the ability of the organization to mobilize the right technological and market skills (including R&D, manufacturing planning and control, and sales and distribution skills). Finally, they can relate to the ability of the project to scale up in a reliable way as the number of adopters grows.

Innovation opportunities related to complex infrastructure projects, such as high-speed trains or nuclear power stations, or to heavily regulated markets, such as healthcare, are notoriously difficult to assess from a feasibility and complexity point of view.

Third, management risks will affect the talent that can be mobilized. They relate in particular to the entrepreneurial skills needed to capture new opportunities (see Chapter 4), the organizational support and the ability to combine both current business activities and new ones (see Chapter 7).

As an illustration, many large incumbents fail to grasp the strategic importance of disruptive innovations (see Chapter 1) as well as to attract, motivate and retain entrepreneurial people (see Chapter 4).

Finally, investment risks will affect the cash flows actually generated. They relate in particular to the actual availability and timing of funding, to the initial profitability of the opportunity (and its ability to be self-financing) and to the cash-out or exit opportunities. As an illustration, many start-ups that could potentially have been successful because there was a sizable market opportunity still become bankrupt because promised funds or expected initial cash-flows are delayed or have been overestimated – for example, because contract negotiations take longer than expected, or because customers pay late.

Furthermore, innovation opportunities that have a long-term perspective and/or global ambitions, for example in the energy, agriculture or airline sectors, also face risks emerging from radical events and externalities. Those include political or military conflicts, economic crises and currency movements as well as strong public resistance resulting from heavily publicized scandals or incidents. As an illustration, many business plans made in the energy sector were completely disrupted by the Chernobyl nuclear power incident in the 1980s, or by the more recent BP oil leak in the Gulf of Mexico. Similarly, the airline sector has to be able to face major disruption such as the events of 9/11 or the more recent volcano eruption in Iceland.

Finally, managers need to be aware of second-order uncertainties. These relate to their inability to "know what they do not know". This means that they need to manage knowledge efficiently in order to minimize what they do not (yet) know what they know. They also need to integrate the variance resulting from what they know they do not know (see below). Finally, they need to build in robustness and value flexibility in order to deal with what they do not (yet) know what they do not know. As an illustration, an international fashion boutique must gather marketing information regarding the future season (leverage knowledge). It also needs to integrate the potential effect of currency movements or economic crises (integrate variance). Finally it needs to build flexibility into the value chain in order to react to unexpected events, such as a sudden demand for a specific model or an unexpected celebrity endorsement.

We shall discuss how to integrate variance in the following section. The issue of developing built-in flexibility will be addressed in Chapter 7.

Integrating variance: know what could happen

There are many ways to analyze, learn and legitimate how the variance emerging from known uncertainties should be integrated in an assessment. We shall discuss here two main types of approach: sensitivity analysis and scenario planning.

Sensitivity analysis starts with an initial assessment of an innovation opportunity (as described earlier in the chapter), based on the most likely or base assumptions. This assessment leads to a base case valuation of the key metrics of the innovation opportunity (such as upfront investments, pay-back time, market potential or net present value).

The second step of a sensitivity analysis is to review all the hypotheses that have been used to value the opportunity (such as market size, market share, prices, marginal costs, fixed costs or timing) and to identify *a priori* the critical parameters. These are the inputs of the valuation that are both highly uncertain – they are the most likely to vary significantly from the base case – and highly relevant – they might affect the outcome of the assessment significantly. As an example, the summer weather is both highly uncertain and highly relevant for bottled drinks and energy firms, but not for electronics or auto companies.

The third step of a sensitivity analysis is to define *a priori* a probability distribution for each of the critical parameters identified. These probability distributions can be very simple (for example, including only a pessimistic and an optimistic value, both with an equal probability) or very complex, based on historical data, simulation models or confidence intervals.

The fourth and final step of a sensitivity analysis is to replicate the initial assessment but change the values of each of the key parameters along the probability distributions that have been defined. This can be done by changing the value of these parameters one by one, creating multiple entry tables or by changing them collectively, defining the worst and best cases. It can also be done by repeatedly changing them simultaneously ("Monte Carlo" simulations).

The main attraction of the sensitivity analysis approach is that it is relatively easy to explain, and that spreadsheet software facilitates the generation of multiple cases. This approach, however, has strong limitations when used to assess innovation opportunities, both in terms of implementation and decision-making.

Sensitivity analysis approaches are quite difficult to implement because they require the definition *a priori* of what are the critical parameters and what is their probability distribution (the second and third steps above). While statistics might be available regarding weather forecasting or unit costs, most real-life innovation opportunities involve critical parameters that are difficult to identify or characterize in advance – such as, for example, technological complexity, future market share or adoption rates.

Moreover, human beings have been shown by numerous psychological studies to tend to overestimate strongly the robustness of their predictions, thus generating dangerous overconfidence. In other words, when defining the probability distributions of the critical parameters, most people will overestimate the accuracy of the base case predictions and underestimate the probability of extreme or "black swan" events.

The famous Western Union quote mentioned earlier in this chapter and the regular financial crises ruining sophisticated and experienced investors provide a clear illustration of this tendency to overconfidence.

A managerial culture of targets and commitment can also limit the ability of managers to consider proactively "what if..." scenarios, because they want to "stick to the plan" and therefore, implicitly or explicitly, take only into account the base case (there is no real "Plan B").

Sensitivity analysis approaches are also challenging in terms of decision-making. It is often difficult to generate results to support decision-making from a sensitivity analysis that are truly meaningful and effective. In many cases, the number of changes in critical parameters that need to be considered can generate so much data that it becomes meaningless. Complex interactions between the critical parameters mean that relevant cases can be missed or lost among many other irrelevant ones. For example, in many cases critical parameters such as market share and adoption rates are not independent variables that can be modified separately.

As a consequence, sensitivity analyses often provide a false sense of confidence, either by generating irrelevant data or by leading to trivial conclusions. This is, for example, the situation when the analysis in fact only shows that, in the worst case assumptions, the project will be less attractive, and that in the best case it will be more attractive.

Sensitivity analyses should therefore be used with caution and only when the critical parameters, their characteristics and their interactions are clearly understood. This is the case, for example, when assessing the business impact of a delay in production, a variation in discount rate or a change in input costs, or when assessing the influence of oil or electricity prices on the profitability of alternative energy sources.

Scenario-based approaches provide an alternative approach, based more on managerial thinking and less on "number crunching". We shall detail these approaches below.

The first step of a scenario approach is to identify, based on an analysis of the business, and not a spreadsheet, what are the main possible evolutions of the environment, of the firm's resources and of its purpose that might affect the innovation opportunity. These main possible evolutions can relate to technologies, markets, regulation or competition, based on the firm's knowledge and intelligence. Each scenario then corresponds to a plausible story regarding these main possible evolutions. The scenario to consider must be both an alternative to the most likely case (not a marginal variation) and have significant potential impact on the opportunity (not an irrelevant situation).

The second step of a scenario approach is to assess, under each of the scenarios considered, the innovation opportunity using the approach described in the previous section. Based on the results of these, the scenarios that could have the most significant impact (positive or negative) on the value of the opportunity are highlighted.

The third step is to analyze, learn and legitimate the business implications of the scenarios highlighted. This concerns, on the one hand, understanding and challenging some necessary conditions for success, if any, and on the other focusing management attention on developing the capability of anticipating, monitoring and reacting to the adverse scenarios identified.

As an illustration, a firm having identified that the entry of a given competitor into its market would have a significant adverse effect can focus on gathering information about the expected moves of this competitor and prepare contingency plans in order to react effectively.

Similarly, a firm understanding that some regulatory evolutions are key to the success of an innovation opportunity will invest resources in tracking the evolution of this regulation and trying to influence it. For example, incumbent telecom firms in many countries successfully lobbied to delay the introduction of "number portability" (the ability to keep the same telephone number when a person switches providers) and "last mile" access (the access to the physical connection to each end-user) because they had understood in advance that these factors had an important impact on their future profitability.

The main attractiveness of scenario-based approaches is that they relate to real-life potential evolutions rather than the manipulations of mathematical parameters. These evolutions can then be interpreted from a business point of view, leveraging the firm's knowledge and expertise. Such approaches also provide guidance regarding what to do if certain events occur, and which evolutions deserve the most management attention.

The main challenges of such an approach are the difficulty of generating relevant scenarios, in particular because of their complexity, and to avoid considering only very likely scenarios and discarding radical or disruptive ones. This can lead to over-confidence and a failure to anticipate critical evolutions.

Again here, the value of the assessment often lies as much in the process itself as in its outcome, as sensitivity and scenario analyses help managers to understand what are the key issues or evolutions on which they should focus. How managers can build-in the flexibility to react to these issues and evolutions will be discussed in Chapter 7.

Key points to take away from Chapter 6

The assessment of an innovation opportunity or "business planning" is not about filling predefined templates. It is first an iterative process, which allows for the analysis of an opportunity, learning about it and legitimizing its potential, if any.

Second, business planning is a communication process, which takes into account the target audience, what are the objectives regarding that audience (for example to enroll, to convince, to excite or to reassure), and what as a result is the best approach and format (a text, a presentation or a "road show", perhaps).

Moreover, from an innovation point of view, business planning is actually not about planning, in the sense of forecasting. Given the uncertain nature of innovation opportunities, business planning is more about, given the information available, identifying whether credible and robust success scenarios can be defined and minimizing the probability of getting things wrong.

While the detailed content and level of complexity of a business plan will be a function of the nature, context and maturity of the innovation opportunity considered, there are four core questions that should always be addressed: "why?", "what?", "who?", and "how much?" Those questions relate to the opportunity potential and to the resources needed to capture it.

The "why?" question is about analyzing, learning and legitimizing whether on the one hand there is a credible opportunity with a significant potential (strategic value), and on the other the organization involved is well-positioned to capture it (competitive differentiation).

The "what?" question is about analyzing, learning and legitimizing whether a robust business model aimed at benefiting from the strategic opportunity identified can actually be designed. It means considering both a potential value proposition (what could be sold to whom) and a potential value chain (how could the value proposition be delivered).

The "who?" question is about analyzing, learning and legitimizing who are the individual and organizational stakeholders that could or should be mobilized to implement the business model under consideration. This includes taking into account the entrepreneurial team itself, the external capabilities needed (including experts and partners) and the governance structure.

Finally the "how much?" question is about analyzing, learning and legitimizing whether the potential opportunity identified is attractive enough to justify the cost of the resources mobilized. This includes understanding the key financial metrics of an opportunity (the market potential, the initial investment and the time horizon) and completing, if appropriate, a careful valuation using, for example, a discounted cash flow approach.

The four core questions outlined above need to be addressed while integrating the underlying uncertainties inherent to innovation opportunities. In particular, the potential implications of "what we know that we do not know" should be integrated with the assessment process – for example, through a sensitivity or (better still) a scenario analysis.

Further reading

Blastland, M. and Dilnot, A. (2007), *The Tiger That Isn't: Seeing Through a World of Numbers*, London, Profile Books.

Christensen, C. M. and Raynor, M. E. (2003), *The Innovator's Solution*, Boston, MA, Harvard Business School Press, Chap. 3.

Copeland, T., Koller, T. and Murrin, J. (2000), *Valuation: Measuring and Managing the Value of Companies*, Hoboken, NJ, John Wiley.

Dalkir, K. (2005), *Knowledge Management in Theory and Practice*, Oxford, Elsevier Butterworth Heinemann.

Davila, T., Epstein, M. J. and Shelton, R. (2006), *Making Innovation Work,* Philadelphia, PA, Wharton School Publishing, pp. 75–79.

Knight, F. (1971), *Risk, Uncertainty and Profit*, University of Chicago Press.

Kotler, P. (2001), *Marketing Management*, Englewood Cliffs, NJ, Prentice-Hall.

Muzyka, D., Birley, S. and Leleux, B. (1996), "Trade-offs in the Investment Decisions of European Venture Capitalists", *Journal of Business Venturing,* Vol. 11, pp. 273–287.

Porter, M. (1996), "What Is Strategy?", *Harvard Business Review*, November/December, pp. 61–78.

Razgaitis, R. (2003), *Dealmaking: Using Real Options and Monte Carlo Analysis*, Hoboken, NJ, John Wiley.

Robock, S. H. and Simmonds, D. (1989), *International Business and Multinational Enterprises*, Homewood, Ill, Richard D. Irwin, Ch. 11.

Wood, M. B. (2004), *Marketing Planning, Principles into Practice*, Englewood Cliffs, NJ, Prentice-Hall.

Managing the innovation pipeline

Short case: changing course

The development of a new drug based on a new molecule, as discussed in Chapter 6, is probably one of the most complex product development processes, and bringing a new drug to market can cost hundreds of millions of euros and has to go through very tight regulatory steps, controlled by both national and international agencies. Finally, the full process has to be completed quickly enough to allow the drug to come to market well before the initial patents expire. Leading pharmaceutical firms have therefore had to develop, over time, the capability of managing product development costs, discipline and speed in an effective way.

During each phase of the development process (Phase I, Phase II, Phase III) firms have to consider whether to pursue the development, stop it or follow an alternative path. Projects that do not meet their initial targets or fall outside the strategic focus of the firm tend to be abandoned.

In the late 1980s, Pfizer was testing a new compound (UK-92-480)[1] that had shown promising results in laboratories and animal tests for the treatment of high blood pressure and angina. There were high expectations regarding the outcome of the "Phase I" clinical trials, as they would

$\rightarrow$

---

determine whether this promising molecule would be developed further.

However the trials did not allow the team to validate the positive effects of the molecule. The results were likely to lead to the interruption of the project, as it had not reached its initial target. In fact, one of Pfizer's main competitors abandoned a similar project at that time.

However, during the trials several male patients also reported a noticeable side-effect. Some even requested to keep the samples or to obtain more pills. Although this was not at all what was intended, the firm had the flexibility to study those side-effects further and to completely reposition the project. It therefore decided to completely reorient the development of the drug.

The new drug was patented in 1996 and was approved for sale in the United States and Europe two years later. It soon became a blockbuster and one of Pfizer's great successes as the first approved effective oral treatment for erectile dysfunction, under the brand name Viagra.

The success of Viagra also prompted Pfizer's competitors to reconsider the project they had abandoned earlier, which ultimately led to the launch in 2003 of Cialis and multiple challenges to Pfizer's patents.

Having developed the capability of identifying a flow of innovation opportunities and assessing which are the best fit with its resources, its environment and its purpose, a firm must ensure that it has the capability to capture those opportunities effectively. This means understanding how projects based on innovation opportunities are different from traditional ones, and what it means regarding their effective and flexible management. Both issues will be addressed below.

Why it's not just project management

There are many project management methodologies available, most of which are based, implicitly or explicitly, on the assumptions "plan–do–check–adjust":

- The context of the project and its key parameters (the "reality") can be defined *a priori* with reasonable precision. The purpose of a project is known in advance, generally based on predefined and measurable performance criteria.
- Given the context, the key parameters and the performance criteria, an optimal plan can be designed and a commitment made to it.
- The project can be managed by tracking any deviations from the initial plan as soon as possible and adjusting resources accordingly.

Hence the process is assumed to be essentially deterministic, as uncertainties are minimized, and linear, as there is a predefined optimal sequence of actions.

As a consequence, the implicit or explicit objective of project management methodologies is in most cases to complete 100 percent of the activities planned and reach exactly the predefined targets for all the projects that have been started.

In such a "plan-and-control" approach, managers can see their job mainly as (i) designing at the outset the optimal plan; (ii) allocating resources and budget accordingly; and (iii) identifying and correcting any deviation from the plan. Seen from that point of view, any project failure is caused either by completely unexpected and unpredictable events or by poor project planning, tracking or follow-up.

While such an approach can be very effective for projects such as building an infrastructure or managing a mass-production process, it is in most cases NOT appropriate when dealing with innovation opportunities.

Projects based on innovation opportunities cannot be assumed to be deterministic nor linear. Projects based on innovation opportunities are not deterministic because of the intrinsic uncertainties of innovation (see Chapter 6). As a consequence, a significant proportion of opportunities that could have been assessed *a priori* as attractive will not deliver the expected results. In many innovative organizations, thousands of potentially attractive ideas typically lead to only hundreds of detailed assessments, to only dozens of projects launched, and ultimately to just a handful of significant business successes.

An important implication of this high attrition rate is that aiming to complete 100 percent of the projects initiated is in fact not a smart objective. On the contrary, the aim should be to detect and stop underperforming projects as soon as possible, and to refocus resources and management attention on the most promising ones.

Second, projects based on innovation opportunities are not linear. Such projects do not in most cases follow a stable sequence of steps or phases. This is because the initial steps of a project and the learning that can be derived from them will affect the context and the key parameters of the subsequent steps of a project (the development process is self-adaptive). Prediction errors, implementation delays, new cognitive models, new threats or opportunities, market changes and so on all mean that what appeared to be an optimal choice in the early stages is not necessarily optimal later in the project.

As examples, a product launch can be rescheduled because of difficulties with initial prototypes; a process might be delayed as a result of intellectual property challenges; or a business model may be redefined following a competitor's entry or a regulatory change. Similarly, a firm can first test which value proposition could generate a significant market potential and positive gross margin, then adjust the proposition based on the results as to which value chain is optimal in the short term to reach break-even, and in the long term to create value.

While such changes of plan can also sometimes occur in the context of traditional projects, they are likely to be much more frequent and significant when dealing with innovation opportunities. They should therefore be integrated with the management of the project itself rather than treated as exceptions or uncontrollable externalities. In other words, one should allow *a priori* for changes in the plan. This means in particular that the implementation of an innovation opportunity should be an iterative process, based on probing and learning rather than a predefined sequence. What is done in a specific step should be adjusted in view of what has been done and learned in the previous steps.

This iterative approach supposes a close collaboration between the teams or departments in charge of the various steps (such as R&D, production, marketing or sales). In particular, one of the most important factors differentiating successful from unsuccessful innovation remains the degree of close collaboration and feedback between product or process design and other corporate functions within the firm, especially manufacturing and marketing.[2]

Such unpredictable and iterative aspects of innovation opportunities mean that mechanistic approaches based on first drawing up a good plan and then sticking to it are not suitable. At the same time, the cost of the resources committed when dealing with innovation opportunities and the resulting risks mean that simple trial-and-error approaches are not acceptable either.

At the organizational level, the challenge for innovative organizations is therefore to achieve speedy ("nimble") execution, combining the effectiveness of a deliberate or "mainstream"[3] approach with the flexibility of emerging or "newstream" ones.

On the one hand, the correct level of a *deliberate* approach must provide the necessary sense of direction, coherence and goal alignment across the organization, mainly through coordination and standardization. The necessary optimization and efficiency is achieved through some level of programming and scheduling, specialization, planning and control, and hierarchy.

On the other hand, the correct level of an *emerging* approach must allow for the flexibility to evolve and to explore options with opportunism and an open mind. It means providing organizational support and accountability to boost entrepreneurship and diversity, to facilitate learning and feedback, experimentation and activism.

Hence it is about finding the right balance between doing things "with perfect replicability, at ever increasing scale and steadily increasing efficiency" and building organizations which are "nimble as change itself".[4] At both team and individual levels, "nimble" execution is about "effectuation" much more than execution. Effectuation is a way of acting and making decisions that has been observed among successful entrepreneurs. It positions the entrepreneur as co-creator and co-developer of opportunities, together with committed stakeholders. In particular, effectual managers consider that:[5]

- The future is fundamentally unpredictable (it cannot be planned), yet it can be controllable through human action (it can be shaped).
- The environment or context is constructible through choice (it is not predefined but part of the project itself).
- The goals and targets of the project are negotiated residuals of stakeholder commitments rather than pre-existing preferences (they evolve).

The ability of an organization to adopt "nimble" approaches and support effectual managers will obviously be a function of its culture and structure (see Chapter 4). It is, however, possible to design and implement specific innovation project management processes to facilitate the necessary balance between effectiveness and flexibility. Such processes will be discussed in the next section.

Adaptative management: stages, portfolios and gates

Having acknowledged that projects based on innovation opportunities will in most cases experience a high failure rate and will evolve significantly over time, managers can implement specific approaches aimed at dealing with these aspects. Approaches include, on the one hand, taking a portfolio perspective, and, on the other, organizing successive stages and gates. Both approaches will be discussed below.

Portfolio management: put your eggs in several baskets

The unpredictability of innovation opportunities means that it is very risky to base the strategy of a firm on a single project, because it is often impossible to guarantee that it will be successful. One solution for firms with sufficient resources is to launch several projects in parallel. In this case they should care about the quality, performance and strategic contribution of the portfolio as a whole, not only of individual projects. This means, on the one hand, assessing the quality of the portfolio, and on the other prioritizing resources across projects to maximize its quality.

An important implication of the portfolio approach is that, when assessing a new opportunity, a firm should consider not only its intrinsic quality (see Chapter 6), but also how it fits with and contributes to the portfolio. A project identified to have some potential might still be rejected, because other more attractive opportunities are available in the portfolio.

A quality portfolio should be sufficiently diversified to minimize the average risk and maximize the overall performance. This is similar to a financial investor diversifying its investments. But a quality portfolio should also be diversified in terms of time horizons, in line with the strategic objectives of the firm. Hence a quality portfolio should balance:

- Investments in short-term projects aimed at extending and defending existing businesses – for example, technical assistance and engineering projects, product and process improvements or line extensions.
- Strategic positions in mid-term projects aimed at developing emerging business opportunities – for example, partnerships, new ventures or new developments.
- Long-term knowledge or competence building initiatives aimed at creating viable future options – for example, corporate research, competence platforms or seed investments.

Keeping the balance between those three time horizons is an important issue. Firms such as some dominant incumbents investing too much in the first time horizon are exposed to the decline of their industries and/or the obsolescence of their business models. Firms such as some ambitious start-ups investing too much in the last time horizon may have a bright future but might not be able to survive long enough to see it.

Finally, a quality portfolio should consider all relevant types and sources of innovation, from internal and incremental R&D-driven process improvements to radical market-driven product innovations and disruptive business models developed through partnerships. Whether and how much each of these types and sources is tapped must be reviewed regularly in line with the strategy (see Chapter 3).

As an example, a firm whose portfolio consists mainly of internal technology-driven projects might be missing opportunities to develop new value propositions or business models. Conversely, a firm focusing on external new business development might be missing opportunities to improve the effectiveness of its processes or its value chain.

We must stress, however, that the diversification in terms of risks, timing and sources of innovation described above must be balanced with the search for scale and scope economies. Projects typically generate fixed costs, in particular in terms of management attention, and resources spread too thinly can often become ineffective. Similarly, a portfolio with many small and unrelated projects will fail to capture significant synergies between those projects. This is the case, for example, in organizations focusing too much on creativity (many ideas are put on the table) and/or unable to "kill" underperforming projects. These organizations end up with a large number of innovation initiatives, but with a low or even a negative impact on their business.

In summary, the portfolio should be diversified enough to mitigate risks and to support both the short- and long-term strategy of the firm. But it should also be coherent enough to be effective and generate synergies.

Focusing on portfolio performance also means that the organization must be able to detect and stop weakly-performing projects early. It must also be able to reallocate the people working on those projects and recreate effective team dynamics (see Chapter 4.). This means in some cases remotivating teams that have been working for months on projects that are then stopped, even if those teams did not necessarily underperform. This is often very difficult, as stopping a project is seen as failing and/or losing face rather than as a learning experience or because performance and reward systems are misaligned.

An important implication for managers is that they must focus their attention and resources on the most promising projects rather than spending a lot of time keeping afloat those that are performing less well.

This is often counterintuitive for managers who are used to plan-and-control approaches, where above-targets projects are essentially left untouched ("on track") while problematic projects receive all their attention ("flashing lights").

A successful innovation portfolio is a portfolio where a few big gold nuggets were uncovered and polished to become beautiful jewels, not a portfolio with many average stones.

As an example, successful venture capitalists often generate high portfolio returns by focusing their attention on one or two stellar projects with exceptional returns and quickly abandoning many other projects. This approach contrasts with typical corporate project portfolios, where all projects are expected to reach acceptable returns, and where projects that have been stopped are synonymous with failure.

Managing innovation projects as a portfolio rather than as individual initiatives allows companies to deal with the uncertain nature of innovation opportunities (they are not deterministic). Acknowledging the fact that such projects do not evolve linearly and must be managed iteratively will be discussed in the next section.

Stage-gate processes

As the context and key parameters of an innovation opportunity evolve, it is critical to step back regularly and reconsider with the relevant stakeholders whether the way the organization is pursuing the opportunity needs to be updated or even completely redefined.

A stage-gate process allows this to be done in a systematic way. In such an approach, a project is separated into various phases (stages) separated by decision points (gates). The objective of such an approach is to replace a single upfront go/no-go investment decision (the project is launched or it is not) by a predefined sequence of more informed and less reversible decisions to commit more resources or not (the project is revisited regularly).

At the end of each stage, a formal review of the opportunity is organized in order to integrate the latest learning and developments. The objective of this review is to decide whether the opportunity should be pursued, stopped, suspended or redefined, and commit or reallocate resources accordingly, given what is known at that point.

The value of a stage-gate process is that it allows reorienting or interrupting a project early. It also allows an increase in the resources committed only after the most recent learning and developments have been taken into account. We must stress that this means more than merely organizing regular progress reviews for each project. As multiple projects are reviewed regularly, and increasing amounts of information and knowledge are mobilized, effective organization of the decision milestones (the gates) is critical. This implies, in particular:

- In terms of timing, balancing the management attention and team preparation costs of organizing multiple and frequent gates against the risks of letting projects drift for too long. Three to five gates are typically organized (sometimes more). These are planned to take place every few weeks or every few months, depending on the speed of evolution of the environment and on the strategic importance of each project.

- In terms of content, defining in advance which key issues must be addressed at each gate, as well as to support and train the project teams gathering the relevant information and presenting their progress.
- In terms of the portfolio, defining by how much the presentation and review of each project must be standardized. The standardization of the review process facilitates the prioritization of projects across the portfolio, but can also lead to too much *a priori* framing of the projects. The risk is that "[when people] only have a hammer, [they] tend to see every problem as a nail" (Abraham Maslow).[6]
- In terms of decision-making, ensuring that the decisions made at each gate are communicated explicitly and unambiguously, and that all the stakeholders involved are committed to those decisions. In particular, whether a project is "killed off" or not, and whether a particular business unit supports a project should be recorded clearly.

The effective organization of the decision milestones (the gates) often requires the allocation of dedicated human resources to take charge of the process itself (process owners), in addition to the people in charge of the individual projects (the project owners) and the people controlling the allocation of resources (the business managers).

As discussed in the previous section, it is important at each gate to review not only the individual projects but also the quality of the portfolio as a whole. Indeed, the decisions taken about each project should take into account what happened to other projects, how it generates relevant learning, how it affects the overall balance of the portfolio, and how it affects shared resources.

Relevant dimensions that should be reviewed at portfolio level include:

- The evolution of the size of the portfolio, in terms of number of projects and total value creation potential.
- How the projects have evolved in terms of perceived accessibility (including technology risks, intellectual property issues

and feasibility studies) and attractiveness (taking into account market size, competitive intensity and commercial risk).
- How much the projects have evolved above or below initial expectations, both in terms of timing and in terms of resources used.

Furthermore, which activities should be completed during the initial stages of each project must also be carefully considered. Indeed, one of the objectives of the stage-gate process is to commit resources only as uncertainty decreases. As a consequence, the stages should be defined in order to deal with critical uncertainties as early as possible and to maximize flexibility and the "learning over investment" ratio.

As an illustration, let us assume that a firm identifies that a specific partnership or piece of regulation is highly uncertain but critical to the success of an innovation opportunity. In this case it should focus the initial stages of the project on resolving the uncertainties, rather than *a priori* developing a prototype or investing in equipment, for example. Similarly, a firm developing a radical innovation for which traditional market research techniques are not robust should focus first on developing a prototype and testing it with a few lead users, rather than systematically investing in wide market surveys.

The aim of each stage should be to decrease as much as possible the uncertainty regarding the innovation opportunity while maximizing the flexibility to make various choices, such as to divest or postpone, as this uncertainty decreases.

This can be done, for example, by identifying in advance which will be the key turning points or critical decision milestones (such as prototype results, market announcements or regulatory decisions) and assess the resulting ranges of options (which decisions could be taken with which impact in terms of resources). This "real option" approach allows the assessment of the value of various potential options (what if...?) and identify different critical path scenario (such as abandonment or deferral, expansion or shrinkage, closure or restart, or switch in specific inputs or outputs). As an illustration, a firm might choose initially to outsource the development of a prototype rather than invest in its own equipment, even if this approach is initially more expensive, because it would allow the flexibility to stop the project more cheaply if it turned out to be less attractive than expected.

An example of the three first stages and gates for managing product innovation opportunities is presented below.

Stage-gate example

In the first stage, a small informal team can complete a short document that is submitted to a local reference person (the process owner). This document should cover the following issues (see Chapter 6):

- "Why?": describe briefly what is the main evidence concerning the target public and what are their needs; which are the main currently available alternatives and their limitations; and how the opportunity fits with the business capabilities, priorities and objectives.
- "What?": describe briefly what are the expected unique features of the concept; who are the potential users; and what are the key technology, intellectual property and implementation issues.
- "Who?": identify the main people (or profiles) expected or needed as project champion, sponsor and experts.
- "How much?": provide a rough estimate of the scale of the expected upfront investment (in full-time equivalents (FTE) and/or cash); benefits (in terms of sales and/or margins); and time to benefit (number of months).

The first gate then consists of periodic (for example, bimonthly) decision meetings organized by the local reference person and representatives of the line management.

These "gate 1" meetings must lead for each proposal submitted to a formal decision, either to reject or to hold the proposal, to transfer/combine/recycle it (stop/hold/revise), or to assign a small team and limited expenses with a deadline to complete a preliminary assessment. This assessment should exploit mainly secondary research and minimal development work, and allow

for the submission of a formal assessment of the innovation
opportunity within four to six weeks.

In the second stage, a small formal team has to prepare a report
and a small presentation, including appendices if needed. This
report consists of an updated and upgraded version of the initial
assessment, plus a consideration of the following issues:

- "Why?": an identification and first quantification of the target segments, the competitive analysis and the key trends as
  well as a perspective on the integration of the project within
  the existing portfolio of initiatives.
- "What?": a proposed potential design of the value proposition for preliminary feedback from the market; a description
  of the key activities and resources needed (value chain), and
  the resulting identified issues, resources and proposed action
  plan.
- "Who?": a proposed project team structure, with expected
  individual levels of involvement; a proposed governance structure; and a perspective on the potential partnerships needed
  (internal and external) and the implications of these.
- "How much?": a breakdown of the main costs and revenues
  over the project timeframe; a preliminary valuation of the
  expected benefits (such as estimated ROI or NPV); and an
  assessment of the main risks and their implications.

The second gate then consists of periodic (for example, quarterly) decision meetings organized by the local reference person
with senior representatives of the line management. During these
"gate 2" meetings, each project is presented by its project team
and sponsors. A formal decision must then be taken to reject,
modify or hold the project, to transfer it or to assign a manager
with a budget and a deadline to complete a detailed assessment
of the proposed investment and prepare an action plan. The formal decision should also involve the confirmation (or not) of the
project team, the main sponsor and the resource people as well
as a formal and motivated feedback to the project team, in particular if the decision is to interrupt or to reorient the project.

In the third stage, the project team must complete a detailed
business plan consisting of a report and/or a presentation with

appendices. This business plan is an updated and upgraded version of the initial plan, addressing in particular:

- "Why?": primary market research (including customer interviews, concept testing and segmentation), an evaluation of the expected market size and its evolution; a profile of the main competitors and a perspective on their expected moves.
- "What?": a proposed initial marketing strategy, with a definition of the value proposition for each target segment; a proposed technology roadmap and operations plan; and a perspective on the resulting supply chain issues.
- "Who?": a project team structure and individual assignments, including a project charter; and a proposed action plan regarding the enrolment of potential partners.
- "How much?": a detailed financial model with a preliminary scenario analysis and, if applicable, a sensitivity analysis.

But even the best of processes are still dependent on the ability of decision-makers to deal with ambiguities and uncertainties, and on the new challenges and issues that each innovation opportunity involves. These aspects will be discussed in the next section.

Why we still get things wrong

Firms dealing with many similar innovation opportunities have over time built a de facto experience of what the issues are that need to be taken care of first in order to decrease uncertainty – such as intellectual property, safety or availability of raw materials. For example, the stage-gate processes implemented by consumer goods firm for their product innovations typically involve both R&D and marketing departments, and start with the definition of a concept, the development of a prototype design, market testing and finally launch and post-launch reviews. Similarly, industrial firms will include stages devoted to production planning and engineering, respectively. In some cases, such as in the pharmaceutical industry, the stages have been standardized at industry level and integrated into the regulations (with Phase I dealing with safety, Phase II dealing with efficacy and side effects, and Phase III dealing with reactions and long-term effects).

However, such implicit or explicit choices regarding the design of the stages (such as always starting with a prototype) might not be adequate when dealing with new types of business opportunities (such as a new business model). In the same way that standardizing the review process at each gate (see above) can frame the decision-making too much, the same risks arise when the stages or phases become routine.

Other key issues regarding the management of a stage-gate process include:

- The ability and sustained motivation to identify opportunities and submit proposals, which can be influenced by the provision of training, rewards and feedback. Many organizations see their stage-gate process quickly "dry up" after the initial launch.
- The ability to deal with potential ambiguities and conflicts about the evolution of the project's scope, regarding, for example, what fits with the core business or what are the corporate priorities, and the project ownership, often split between project champions and line managers.
- Paperwork overflow (leading sometimes to "analysis paralysis", or the inability to make decisions without looking for more data) and the practical organization of the gate reviews, of virtual collaboration, of knowledge management and of the networking within and across projects.
- The management of external projects, and in particular the alignment of partners with corporate processes, cultures and systems (see the open innovation discussion in Chapter 4).
- The management of stopped projects in terms of feedback, resource reallocation and lessons learned, as well as the sustained tolerance of a relatively high failure rate.

In summary, a stage-gate process cannot be implemented in an organization simply based on existing templates or software. Its legitimacy across the organization and its integration within the corporate culture, routines and values are key success factors, which are more important than deciding whether software X or template Y should be used. In particular, the role of training (for example, about how to make an assessment and how to present it to managers) and the motivation of local managers and reference people (the process owners) should not be underestimated.

But even an effective stage-gate process does not completely remove the ambiguities and uncertainties of innovations. It can therefore fail if the wrong decisions are taken, either regarding what happens at a given gate or what is expected from a given stage. The common decision-making issues and biases that might affect such decisions include:[7]

- The confirmation bias: when facing a difficult or ambiguous decision, human beings are likely to hear only what agrees with their initial hypothesis and deny alternative perspectives. The ability of the management to hear dissenting voices and question them a priori is therefore key.
- The sunk cost fallacy: most managers are uncomfortable with stopping a project, especially when significant financial resources have already been injected. Stopping a project only means recognizing that the money has in fact already been lost, but some managers might see it as losing faces.
- The "boiling frog" process (if you "cook" a frog in a bowl of water slowly enough, it will apparently die from the heat rather than jump out), or escalation of commitment: the success of a project is always seen as "just round the corner" and incremental resources continue to be spent without significant progress being made. The organization eventually engages in levels of investment that would have never been accepted in advance.
- The "anchoring" process: human subjectivity tends to take what is seen first as being the most important. But in decision-making processes, the first impression is not always the best. Managers should be able to question initial estimates and review their mental models and targets.

But knowing that these biases exist, and recognizing that they exist not only when *other* people make decisions, does not provide a formula for dealing with them. Acknowledging them is, however, a step in the right direction.

Key points to take away from Chapter 7

Standard project management techniques based on "plan–do–check" approaches are ill-suited to the capture of innovation opportunities, because of the intrinsic uncertainties linked to such projects and because of their iterative nature. What is needed is an execution approach that is both effective and flexible ("nimble").

The uncertain nature of innovation opportunities means that a significant share of these opportunities will not be implemented successfully. It is therefore critical to deal with this expected failure rate, in terms of decision-making, resource reallocation, staff motivation and learning.

The iterative nature of innovation opportunities means that some key aspects of such opportunities and how much we know about them will change significantly as the opportunities are implemented. It is therefore critical to be able to question regularly the initial assumptions upon which the implementation is based and to reorient the projects accordingly.

"Nimble" execution means finding a balance between the efficiency needed to save time and resources, and the flexibility needed to adjust proactively as uncertainties are resolved and key elements of the projects evolve.

The first aspect of a "nimble" execution approach is to manage the various innovation opportunities as a project portfolio, in order to mitigate risks, capture synergies and balance opportunities in terms of resources and time horizons.

The second aspect of such an approach is, on the one hand, to structure the projects in phases with decreasing uncertainties and increasing commitments, and on the other hand to integrate specific decision-points in the process. After each phase (stage), a decision point (gate) should be set up in order to capitalize on what has been learned and reorient the implementation of the project accordingly.

A stage-gate process must be fine-tuned carefully to the context and culture of the organization implementing it, in particular in terms of the reporting and decision-making process (the format and management of the gates meetings) and in terms of human resource management (keeping the people behind the projects motivated and effective).

Even when effective stage-gate processes are implemented, organizations remain vulnerable to individual and collective decision-making biases that are rooted in human psychology, culture and mental models. Being aware of these biases is the first step in the right direction. When managing innovations, what matters is not always to be right, but to be wrong less often than others and to adjust more quickly when it happens.

Further reading

Baghai, M., Coley, S. and White, D. (1999), *The Alchemy of Growth: Kickstarting and Sustaining Growth in Your Company*, New York, The Free Press.

Chesbrough, H. (2004), "Managing Open Innovation", *Research-Technology Management*, Vol. 47, pp. 23–26.

Cooper, G. C. (2002), "Optimizing the Stage–Gate Process", *Research-Technology Management*, Vol. 45, pp. 43–49.

Hamel, G. and Bryan, L. (2008), "Innovative Management", *McKinsey Quarterly*, Vol. 2008/1, pp. 25–35.

Haspeslagh, P. (1982), "Portfolio Planning: Uses and Limits", *Harvard Business Review*, January/February, pp. 58–73.

Horn, J. T., Lovallo, D. P. and Viguerie, S. P. (2006), "Learning to Let Go: Making Better Exit Decisions", *McKinsey Quarterly,* Vol. 2006/2, pp. 65–75.

McKinsey & Company (1999), *Breaking Down Corporate Boundaries to Unleash Innovation*, New York, McKinsey & Company.

Mikkola, J. H. (2001), "Portfolio Management of R&D Projects", *Technovation*, Vol. 21, pp. 423–435.

Mintzberg, H. and Waters, J. A. (1985), "Of Strategy: Deliberate and Emergent", *Strategic Management Journal*, Vol. 63, pp. 257–272.

Razgaitis, R. (2003), *Dealmaking: Using Real Options and Monte Carlo Analysis* Hoboken, NJ, John Wiley.

Sarasvathy, S. (2008), *Effectuation: Elements of Entrepreneurial Expertise*, Cheltenham, Edward Elgar.

Schilling, M. A. (2006), *Strategic Management of Technological Innovation,* 2nd edn, New York, McGraw-Hill, pp.137–138.

Tidd, J., Bessant, D. and Pavitt, K. (2001), *Managing Innovation*, 2nd edn, Hoboken, NJ, John Wiley, pp. 138–147 and 367.

Van de Ven, A. H., Polley, D. E., Garud, R. and Venkataraman, S. (2008), *The Innovation Journey,* Oxford University Press, p. 185.

Synthesis of Part II

The first part of this book focused on understanding what innovation means and why it matters from a business point of view. This second part discussed the capabilities an organization must build and maintain if it wants to be able to manage innovation effectively.

The first capability is to realize that innovation is a means to achieve a company's strategic objectives, not an end *per se*. It is therefore critical to be able to define and share across the organization an effective innovation strategy. This is about having a sense of direction and steering the "innovation engine" of the organization accordingly.

The second capability is to recognize that innovations are above all made by people and teams, not processes or systems. It is therefore a key success factor to be able to develop and leverage internal and external entrepreneurial resources. These must allow the organization to identify and assess innovation opportunities, as well as to build coalitions in order to take advantage of such chances. This is about energizing the "innovation engine" at individual, team, organization and ecosystem levels.

The next three capabilities are about managing effectively the corporate entrepreneurial process, which is the ability of the organization to identify opportunities, to assess which are the most attractive given the strategic objectives, and to mobilize the resources needed to capture them in an effective but flexible way. This is about accelerating the "innovation engine" of the organization.

All this means that building an innovative organization is about much more than boosting creativity, sharing new ideas or increasing the R&D budget. What should be on the corporate agenda of an innovative firm is to evaluate and improve continuously each of the five organizational capabilities discussed above, knowing where the weakest links are and dealing with them proactively.

Conclusion

One common innovation management delusion is to start the innovation journey with the image of genius inventors in mind who became rich and famous by coming up with revolutionary products. Accordingly, managers try to detect and leverage latent heroes in their organizations by boosting creativity, increasing R&D budgets or setting up idea portals on the corporate intranet. Once all these initiatives have failed to have a significant impact on corporate performance, the managers realize that innovation management is more challenging than they thought, and that other issues have to be dealt with.

Another common delusion is to see innovation as a fad, part of a long sequence of business concepts that come and go like quality circles, business process reengineering, or online intermediaries. Accordingly, some managers tolerate and comply with innovation initiatives in their organizations but expect them to fade away as the next business cycle emerges or as a new business guru publishes his or her book.

The final delusion is to see innovations as the solution to all problems, whether it be lack of competitiveness, job losses or economic stagnation. In this case, innovation becomes an end *per se*, a quantitative objective where more is always better. The level of R&D spending, the number of new businesses created, the number of new ideas and product launches or the number of patents become holy criteria to be pursued by all in all circumstances.

The main aim of this book was to help active and future managers to avoid falling in those traps. Whatever your business, whatever your sector, innovation does matter. But the challenge is to understand what it means, how much it matters and how to deal with it. Innovation matters because innovations have become more frequent and more prevalent. From banking to consumer goods, innovations are not (and have never been) the preserve of high-tech firms and scientists. When innovation opportunities are created and shared across the globe at the speed of light, they can no longer be assumed to be only accidental events, occurring between calm periods when businesses can be managed as usual. Innovation management is an issue that should be on the agenda of all firms, small and large, old and new.

What innovation means is much more than coming up with new ideas generated out of the blue. Innovations are in fact often not completely new. They are based largely on combinations of existing expertises and new perspectives as well as on the reinvention or adaptation of old concepts. Innovations are also more about storming resources and people than just storming brains. As many inventors who failed to capture the benefits of their innovations have discovered, innovation is (to paraphrase the famous inventor, Thomas Edison) about 1 percent inspiration and 99 percent perspiration. Innovation is about understanding what drives change and how to make it happen.

Managing innovations also means much more than launching revolutionary new products. Most innovations relate to processes (*how* we make) rather than products (*what* we make) and are about many small changes (incremental) rather than revolutions (radical). Even firms built on "big ideas" such as, initially, Ford or more recently Apple, can succeed only if they can follow up those revolutionary concepts with many improvements not only to the concepts themselves but also to their implementation. An average opportunity well executed is worth a hundred great ideas poorly delivered.

Knowing what innovation actually means from a business point of view and why it matters is the first step. What matters after that is to know what to do about it and how to detect and deal with the resulting threats and opportunities.

Since Schumpeter we know that innovations are the products of entrepreneurs, who are able to see opportunities, decide which ones to pursue and mobilize resource accordingly. Innovative firms must therefore develop the capability of acting as corporate entrepreneurs, identifying opportunities, assessing and exploiting them.

Most people are not spontaneously entrepreneurial. What matters is to identify and support the cultural and structural levers that will increase the will and skill of an organization, its people, its teams and its partners to act as entrepreneurs. Conversely, an entrepreneurial opportunity that is good for one firm might not be for another. Facilitating and nurturing the entrepreneurial process makes sense only if the organization can steer it, because it has a clear and shared vision of what it wants to achieve, whether in terms of growth, profitability or stakeholder expectations.

Obviously, developing such innovation management capabilities in a given firm can present significant human and technical challenges and has to be fine-tuned to its specific characteristics and context. Managing innovation in a small, high-tech start-up, in an international service company or in a large industrial corporation also means dealing with specific technical issues that go beyond the scope of this book.

The good news is that our experience with many executives from various sectors is that the core issues addressed in this book are indeed valid across firms and sectors. Even if the scale, the specific technology used or the time horizon change, the underlying issues and dilemmas remain, whether in terms of dealing with risk and change, or combining efficiency and flexibility.

The next piece of good news is that if it were easy to manage innovation, everybody would do it and there would be no room left for competitive differentiation. The fact that innovation management is not a commoditized capability makes it even more valuable.

The third piece of good news is that management is not a science like physics or engineering. There is no definitive right answer that a manager needs to find or otherwise fail. In innovation, what we know is that even the best managers will sometimes get it wrong. What matters is that they get it wrong less and less often than their competitors, and that they are able to recognize and act quickly to correct their mistakes.

Chapter 1

1. Evans, H., Buckland, G. and Lefer, D. (2004), *They Made America: From the Steam Engine to the Search Engine: Two Centuries of Innovators*, New York, Little, Brown, p. 11.
2. Van de Ven, A. H. (1986), "Central Problems in the Management of Innovation", *Management Science*, Vol. 32, pp. 590–607.
3. Stinchcombe, A. L. (1965), "Social Structure and Organizations", in J. G. March (ed.), *Handbook of Organizations*, Skokie, IL, Rand McNally, pp. 153–193.
4. Quoted by F. M. Scherer (1965), "Invention and Innovation in the Watt–Boulton Steam-Engine Venture", *Technology and Culture*, Vol. 6, pp. 165–187.
5. Berkun, S. (2007), *The Myths of Innovation*, Sebastopol, CA, O'Reilly Media, p. 8.
6. Fagerberg, J. (2005), *The Oxford Handbook of Innovation*, Oxford University Press, p. 21.
7. BCG (Boston Consulting Group) (2009), *Innovation 2009: Making Hard Decisions in the Downturn*, Boston, MA, BCG, p. 11.
8. Hamel, G. and Breen, B. (2009), *The Future of Management*, Boston, MA, Harvard Business School Press, p. 96.
9. Rogers, E. M. (1995), *Diffusion of Innovations*, New York, Free Press, pp. 376–383.
10. Hard, M. and Knie, A. (2001), "The Cultural Dimension of Technology Management: Lessons from the History of the Automobile", *Technology Analysis & Strategic Management*, Vol. 13, pp. 91–103.
11. Hall, B. H. (2005), "Innovation and Diffusion" in J. Fagerberg, Mowery, D. C. and Nelson, R. R. (eds) *The Oxford Innovation Handbook,* Oxford University Press, pp. 462–481.
12. Pavitt, K. (2005), "Innovation Process", in J. Fagerberg Mowery, D. C. and Nelson, R. R. (eds), *The Oxford Innovation Handbook,* Oxford University Press, p. 108.
13. Hard and Knie, as Nt 10.
14. Rogers, E. M. (1995), *Diffusion of Innovation*, 4th edn. New York, Free Press, pp. 252–281.
15. Moore, G. A. (2002), *Crossing the Chasm: Marketing and Selling High-Tech Products to Mainstream Customers*, New York, Harper Paperbacks, p. xiv.

16. Utterback, J. M. (1996), *Mastering the Dynamics of Innovation*, Boston, MA, Harvard Business School Press, pp. 23–56.
17. Christensen, C. (1997), *The Innovator's Dilemma: When New Technologies Cause Great Firms to Fail*, Harvard Business School Press.
18. Christensen, C. M. and Raynor, M. E. (2003), *The Innovator's Solution* Boston, MA, Harvard Business School Press, p. 33.
19. OECD/Eurostat (2005), *Guidelines for Collecting and Interpreting Innovation Data — The Oslo Manual*, 3rd edn, Paris, OECD, p. 124.
20. OECD/Eurostat, as Nt 19.
21. J. M. Utterback (1996), *Mastering the Dynamics of Innovation*, Boston, MA, Harvard Business School Press, p. xvii.

Chapter 2

1. OECD (2009), *Innovation in Firms: A Microeconomic Perspective*, Paris, OECD, pp. 29–30.
2. Beinhocker, E. D. (1997), "Strategy at the Edge of Chaos", *McKinsey Quarterly*, 1997/1, pp. 24–39.
3. Beinhocker, as Nt 2.
4. De Wit, B. and Meyer, R. (2003), *Strategy: Process, Content, Context*, Florence, KY, Cengage Learning EMEA, Ch. 10.
5. Becker, W. M. and Freeman, V. M. (2006), "Going from Global Trends to Corporate Strategy", *McKinsey Quarterly,* 2006/3, pp. 17–27.
6. Van de Ven, A. H., Polley, D. E., Garud, R. and Venkataraman, S. (2008), *The Innovation Journey*, Oxford University Press, p. 117.
7. Shane, S. and Venkataraman, S. (2000), "The Promise of Entrepreneurship as a Field of Research", *Academy of Management Review*, Vol. 25, pp. 217–226.
8. Mintzberg, H. (1973), "Strategy Making in Three Modes", *California Management Review*, Vol. 16, pp. 44–53.

Chapter 3

1. Hamel, G. and Prahalad, C. K. (1994), "Competing for the Future", *Harvard Business Review*, July/August, pp. 122–128.
2. Levitt, B. and March, J. G. (1988), "Organizational Learning", *Annual Review of Sociology,* Vol. 14, pp. 319–340.
3. Hambrick, D. C., MacMillan, I. C. and Day, D. L. (1982), "Strategic Attributes and Performance in the BCG Matrix", *Academy of Management Journal*, Vol. 25, pp. 510–531.
4. Gluck, F. W., Kaufman, S. P., Walleck, A. S., McLeod, K. and Stuckey, J. (2000), "Thinking Strategically", *McKinsey Quarterly*, June.
5. Zook, C. (2007), "Finding Your Next Core Business", *Harvard Business Review*, April, pp. 66–75.

6. OECD (2009), *Innovation in Firms: A Microeconomic Perspective*. Paris, OECD, p. 34.
7. Hamel, G. (1998), "Strategy Innovation and the Quest for Value", MIT *Sloan Management Review*, Vol. 39, pp. 7–14.
8. Markides, C. C. (1999), *All the Right Moves: A Guide to Crafting Breakthrough Strategy*, Boston, MA, Harvard Business School Press, p. 19.
9. Chan, K. and Mauborgne, R. (1999), "Strategy, Value Innovation and the Knowledge Economy", *MIT Sloan Management Review*, Vol. 40, pp. 41–54.
10. Govindarajan, V. and Trimble, Ch. (2007), *10 Rules for Strategic Innovators: From Idea to Execution*, Boston, MA, Harvard Business School Press, p. 183.
11. Penrose, E. T. (1959), *The Theory of the Growth of the Firm*, Hoboken, NJ, John Wiley.
12. Teece, D. and Pisano, G. (1994), "The Dynamic Capabilities of Firms: An Introduction", *Industrial and Corporate Change*, Vol. 3, pp. 537–556.
13. Quoted by J. Rigsby and G. Greco (2002), *Mastering Strategy,* New York, McGraw-Hill, back cover.

Chapter 4

1. This is a disguised version of a real-life case.
2. Mintzberg, H. (1973), "Strategy Making in Three Modes", *California Management Review*, Vol. 16, pp. 44–53.
3. Stevenson, H. H. and Jarillo, J. C. (1990), "A Paradigm of Entrepreneurship: Entrepreneurial Management", *Strategic Management Journal*, Vol. 11, pp. 17–27.
4. Both quoted by *The Economist* (2009) "Global heroes: A special report on entrepreneurship", March 14, 2009, p. 2
5. Shane, S. (2008), *The Illusions of Entrepreneurship*, New Haven, CT, Yale University Press, p. 45.
6. Burgelman, R. A. (1983), "A Process Model of Internal Corporate Venturing in the Diversified Major Firm", *Administrative Science Quarterly*, Vol. 28, pp. 223–244.
7. Sharma, P. and Chrisman, J. J. (1999), "Toward a Reconciliation of the Definitional Issues in the Field of Corporate Entrepreneurship", *Entrepreneurship Theory and Practice*, Vol. 23, pp. 11–26.
8. Pinchot, G. (1985), *Intrapreneuring: Why You Don't Have to Leave the Corporation to Become an Entrepreneur*, New York, Harper & Row.
9. Maitland, A. (2007), "Dyslexic Entrepreneurs" ; available at http://www.cass.city.ac.uk/ media/stories/Dyslexia.html.
10. Gartner, W. B. (1988), "Who Is an Entrepreneur is the Wrong question", *American Journal of Small Business*, Vol. 12, pp. 11–32.

11. Gartner, W. B. (1985), "A Conceptual Framework for Describing the Phenomenon of New Venture Creation", *Academy of Management Review*, Vol. 10, pp. 696–706.
12. Ajzen, I. (2001), "Nature and Operation of Attitudes", *Annual Review of Psychology*, Vol. 52, pp. 27–58.
13. West, M. A. (2002), "Sparkling Fountains or Stagnant Ponds", *Applied Psychology: An International Review*, Vol. 5, pp. 355–524.
14. Quoted by J. B. Bickel (2007), "Turning Intellectual Capital into Leadership Capital: Why and How Psychiatrists Can Take the Lead", *Academic Psychiatry*, Vol. 31, pp. 1–4.
15. IBM (2006), *Expanding the Innovation: The Global CEO Study 2006*, Armonk, NY, IBM Business Consulting Services, p. 30.
16. Adapted from G. T. Lumpkin and G. G. Dess (2001), "Linking Two Dimensions of Entrepreneurial Orientation to Firm Performance", *Journal of Business Venturing*, Vol.16, pp. 429–51.
17. Tushman, M. L. and O'Reilly, C. A. (1996), "Ambidextrous Organizations: Managing Evolutionary and Revolutionary Changes", *California Management Review*, Vol. 38 pp. 8–31.
18. Teece, D. (2009), *Dynamic Capabilities and Strategic Management*, Oxford University Press, p. 227.
19. Teece, as Nt 18.
20. Pohle, G. and Wunker, S. (2007), *Innovating in Your Own Terms*. New York, IBM Institute for Business Value.
21. OECD (2009), *Innovation in Firms: A Microeconomic Perspective*, Paris, OECD.
22. Granovetter, M. (1985), "Economics Action and Social Structure: The Problem of Embeddedness", *American Journal of Sociology*, Vol. 91, pp. 481–510.

Chapter 5

1. This is a disguised version of a real-life case.
2. Mathieu, A. and van Pottelsberghe de la Potterie, B. (2008), "A Note on the Drivers of R&D Intensity", *CEPR Discussion Paper No.6684.*
3. Prahalad, C. K. and Hamel, G. (1990), "The Core Competencies of the Corporation", *Harvard Business Review*, May/June, p. 81.
4. Smith, K. (2005), "Measuring Innovation", in J. Fagerberg Mowery, D. C. and Nelson, R. R. (eds), *The Oxford Innovation Handbook* Oxford University Press.
5. De Geus, A. (1997), *The Living Company*, Boston, MA, Harvard Business School Press, p. 133.
6. Diamond, J. (1998), *Guns, Germs and Steel: A Short History of Everybody for the Last 13000 Years*, New York, Vintage, pp. 234–236.
7. OECD (2009), *Innovation in Firms: A Microeconomic Perspective*, Paris, OECD, p. 158.

8. Dushnitsky, G. and Lenox, M. J. (2005), "When Do Incumbents Learn From Entrepreneurial Ventures?", *Research Policy*, Vol. 34, pp. 615–639.
9. Maula, M. V. J. (2007), "Corporate Venture Capital as a Strategic Tool for Corporations" in H. Landström (ed.) *Handbook of Research on Venture Capital*, Cheltenham, Edward Elgar, pp. 371–392.
10. Chesbrough, H. (2000), "Designing Corporate Ventures in the Shadow of Private Venture Capital", *California Management Review*, Vol. 42, pp. 31–49.
11. Dushnitsky, G. and Lenox, M. J. (2005), "When Do Firms Undertake R&D by Investing in New Ventures?", *Strategic Management Journal*, Vol. 26, pp. 947–965.

Chapter 6

1. This is a disguised version of a real-life case.
2. Van de Ven, A. H., Polley, D. E., Garud, R. and Venkataraman, S. (2008), *The Innovation Journey*, Oxford University Press, p. 23.
3. Copeland, T., Koller, T. and Murrin, J. (2000), *Valuation: Measuring and Managing the Value of Companies*, Hoboken, NJ, John Wiley.
4. Quoted by Y. Sheffi (2004), "RFID and the Innovation Cycle", *International Journal of Logistics Management*, Vol. 15, pp. 1–10.

Chapter 7

1. The key facts of this brief history have been adapted from A. Keith (2000), "The Economics of Viagra", *Health Affairs,* Vol. 19, No. 2, pp. 147–157; and "Sildenafil", available at en.wikipedia.org/wiki/Viagra.
2. Rothwell, R. (1992), "Successful Industrial Innovations: Critical Success Factors for the 1990s", *Research Policy*, Vol. 22, No. 3, 221–239; quoted by J. Fagerberg Mowery, D. C. and Nelson, R. R. (eds), *The Oxford Innovation Handbook* Oxford University Press, p. 90.
3. O'Reilly, C. A. III and Tushman, M. L. (2004), "The Ambidextrous Organization", *Harvard Business Review*, April, pp. 74–81.
4. Hamel, G. and Breen, B. (2009), *The Future of Management*, Boston, MA, Harvard Business School Press, p. 41.
5. Read, S., Song, M. and Smit, W. (2009), "A Meta-analytic Review of Effectuation and Venture Performance" *Journal of Business Venturing*, Vol. 24, pp. 573–587.
6. Quoted by J. R. Rowling (2002), *Heading towards Excellence*, Stoke-on-Trent, Trentham Books, p. 42.
7. Adapted from T. Gilovich, D. Griffin and D. Kahneman (2002), *Heuristics and Biases: The Psychology of Intuitive Judgment*, Cambridge University Press.

adopter, *see* users
adoption, 10–11, 39, 52, 149–51, 155
  chasm, 16, 47
  stages, 16
airline industry, 43, 56
ambidextrous organization, 93, 107,
  114
Apple, 10, 46, 105, 157
automobile industry, 26, 54, 57, 59
autonomy, 82, 91

banking, 50–1, 57
Baumol, William, 73
BCG matrix, 48
Beinhocker, Eric, 29
BMW, 54, 59
break-even, 86
BRIC countries, 31, 114
bureaucracy, 76, 90, 92, 98, 117
business model, 20, 44, 58, 68, 148,
  154
business planning, 124, 143, 174–5,
  190
business strategy, 44, 50, 67

cell phones, 3–5
change, 6–7, 9–12, 23, 34, 39, 93,
  199
client management, *see* sales and
  marketing
Coca-Cola, 46, 66, 92
collaborations, *see* partnerships
combinations, 8–9, 199
competency traps, 46, 90
competitive advantages, 29, 147,
  151, 153
competitive environment, 29, 32, 57,
  61, 148
competitive pressures, 32

continuous improvements, 22–3, 87,
  93
copyright, 128
core business, 45, 47
core rigidities, 49
corporate entrepreneurs, 35, 72, 75
corporate entrepreneurship, 73, 196
corporate strategy, 44, 67
corporate venture capital, 97, 134–7,
  139
corporate venturing, 94–5, 108
cost of capital, 164–6
creativity, 5, 9–10, 39, 91
cultural ambience, 14

decision-making biases, 193
decision-making process, 83, 90
deliberate/emerging approaches, 181
Dell, 66, 115
design, 55–6, 156
diffusion, 11–12, 39
  drivers of, 12–14
  managing, 14–15
  process, 15–17
discount rate, *see* cost of capital
discounted cash-flow, *see* net present
  value
discovery, 6, 9
disruption, 11–13, 39
disruptive innovation, 17–18, 33, 69
diversification, 48, 135, 183–4
dominant design, 16–17, 53
drivers of entrepreneurial intentions,
  77–9
  attitudes towards entrepreneurial
    behavior, 78–9
  perceived behavioral control, 78–9
  subjective norms, 78–9
Drucker, Peter, 64, 73

Edison, Thomas, 10
"effectuation", 182
electric cars , 26–7
enrollment, 75–6, 144
entrepreneurial behaviors, 75–6,
    78
entrepreneurial orientation, 91
entrepreneurial process, 35–6
entrepreneurial traits, 74–5
entrepreneurs, 72–3, 199
entrepreneurship, 72, 75
Eureka moment, 5, 8
exploitation/exploration, 23, 34,
    45–7, 49, 67, 87–8, 93–4, 107,
    113
external ventures, 96–7, 135
Exxon, 46, 60

fears of innovation, 10–11
Ferrero, 46
final value, 164–6
financials, 160–2
first mover/follower, 50, 52–5, 68
flexibility, 29, 181, 188, 194, 200
Fortune 500, 32

gatekeepers, 110, 124, 138
GE–McKinsey matrix, 48
General Electric, 10
globalization, 30–1
Google, 84, 105, 154
governance, 83, 90, 158–9
group task orientation, 82

ideas, 5–6, 86, 122, 199
incremental innovation, 22–3, 39,
    88, 120
incumbents, 18, 39
industry trends, 32–4
information technology, 20, 57
innovation capabilities, 35–6, 63–4,
    200
innovation champions, 71, 90
innovation culture, 89–91, 106–7,
    137
innovation ecosystem, 99–102, 104,
    108

innovation infrastructure, 99–101
innovation metrics, 85–8
innovation networks, 102–3
innovation opportunities, 112, 121,
    144, 146–7, 199
innovation project leader, 83–4
innovation project staffing, 107
innovation support mechanisms,
    101–2
innovative teams, 80–4, 158–9
    diversity, 82
    dynamics, 83
intellectual capital, 125
intellectual property, 63, 87, 90,
    99–100, 104, 106, 110, 125,
    128–31, 137
internal ventures, 95–6
    board, 96
international competition, 31
internationalization, 30–1, 114–15
invention, 7–9
inventiveness, 6, 39
investment, 162
iterative process, 181, 194

knowledge, 125, 127
    absorption, 126
    explicit/tacit, 125–6
    management, 125–7
Kodak, 9

lead users, 120
leadership, 83–4, 89
learning, 89, 125, 180–1, 186, 188
licensing, 104, 129
lock-in, 54
locus of control, 74
L'Oréal, 66
LVMH, 66

mainstream/newstream, 47, 181
make or buy, 56, 159
market failures, 100–1
market potential, 149–50, 161
market research, 120, 123
mechanistic/organic approach, 93,
    125–6

mental models, 118
Microsoft, 67

needs, 148
negotiation, 105–6, 129, 136
net present value, 163–6
network externalities, 53, 100–1
new business development, 87–8, 95, 98, 110
new innovation mindsets, 87–8, 98
new ventures, 73
new world, 29
newness, 6–9, 11, 93
"nimble" execution, 182, 194
Nintendo, 67
Nokia, 46, 66
not invented here, 83, 91

offer, *see* value propositions
open innovation, 104–6, 108
open source, 131
operations, 56, 156
organizational capabilities, 35–6, 196, 200

partnerships, 86, 90, 102–3, 125, 158–9
patents, 41, 54, 79, 86, 120, 128–31, 151
  infringement, 130–1
pay-back, 163
Penrose, Edith, 63
Pfizer, 177–8
picking winners, 100
Porter, Michael: five forces, 61
portfolio, 47–9, 136, 183–5, 187
prioritization, 143–4
process innovation, 21–2, 39
Procter & Gamble, 65, 105
product development, 19, 177, 191
product innovation, 19, 21, 39
product/market positioning, 20, 44, 50–1, 57, 67
project management, 81, 179, 194
purpose, *see* strategic objective
push/pull, 119–21, 132, 138

R&D, 33, 55, 91, 110, 112–15, 123, 137
  organization, 113–15
  spending, 86–7, 113, 198
  strategy, 113
radical innovation, 22–3, 39, 57, 88, 119
real options, 188
regulation, 14–15, 99, 169
Renault, 54, 59
resources, *see* strategic assets
risk-taking, 65, 79, 88–91, 101
Rogers, Everett, 12–13
role model, 74, 78, 107

sales and marketing, 55, 57, 157
scenario planning, 173–4
Schumpeter, Joseph, 8
scouting, 105
search behaviors, 122–4
  institutional, 123–4
  problematic, 122–3
  slack, 84–5, 123
search engines, 53
sensitivity analysis, 171–3
service innovation, 20
Siemens, 18
size, 92–3
socio-economic environment, 29, 61–2, 99–102, 117–18, 152
sources of innovation , 112, 116–18, 138, 184
SouthWest Airlines, 66
spin-off, 95, 97, 104
stage-gate, 186–91, 194
stakeholders, 33, 64–6, 145, 153
state interventions, 100–1
stopping a project, 185, 189, 192
strategic assets, 48–9, 52, 63–4, 92, 129, 152–3
strategic commitments, 54, 92
strategic fit, 60, 151
strategic gap analysis, 65
strategic innovation, 58–60, 68
strategic objective, 35, 65, 67–8, 136, 153
strategic posture, 66–7, 81

strategic priority, 60–1, 66
strategic value creation levers, 45,
    147, 161
strategic vision, 35, 64–5
stretch *v.* stress, 84–5
support functions, 55, 57, 157
sustainable competitive advantages,
    29–30
SWOT, 61, 63

technology clusters, 103
technology platforms, 113
Telefonica, 66
3M, 65, 77, 84, 157
tolerance for failure, 89
Toyota, 59
trends, 28–32, 40

uncertainties, 34, 53, 59, 65, 88,
    106, 145, 167–70, 180,
    188
  second-order, 170
  sources of, 168–70
users, 149–51, 161

valuation, 167
value chain, 21, 44, 50, 55–7, 68,
    154–6, 161
value gap analysis, 132–3, 138
value propositions, 19, 50, 55, 59,
    133, 148, 154–5, 161
venture capitalists, 163, 167,
    185
Vodaphone, 49
Volkswagen, 66